Synthesis Lectures on Emerging Engineering Technologies

This series publishes short books on current engineering technologies that are gaining prominence, as well as promising technologies that are being developed, for an audience of researchers, advanced students, engineers and other professionals, and entrepreneurs.

Ana Cristina Faria Ribeiro · A. K. Haghi

Smart Water Resource Management

A Practical Introduction

 Springer

Ana Cristina Faria Ribeiro
Department of Chemistry
University of Coimbra
Coimbra, Portugal

A. K. Haghi
Department of Chemistry
Institute of Molecular Sciences,
University of Coimbra
Coimbra, Portugal

ISSN 2381-1412   ISSN 2381-1439 (electronic)
Synthesis Lectures on Emerging Engineering Technologies
ISBN 978-3-031-60798-1   ISBN 978-3-031-60799-8 (eBook)
https://doi.org/10.1007/978-3-031-60799-8

This Springer imprint is published by the registered company Springer Nature Switzerland AG
The registered company address is: Gewerbestrasse 11, 6330 Cham, Switzerland

If disposing of this product, please recycle the paper.

Preface

It's important to pay the world today serious attention to environmental engineering and smart water systems for water reliability, economic efficiency, and sustainable development. The topic of loss control in water facilities is of particular importance. The value of water production in the 22nd century will be many times higher compared to the production of water in the 21st century. The reason for the increase in the cost of water production is the inevitable implementation of consumption management programs in the future. In other words, water production will be equivalent to the most expensive production resources in the future. The effective factors in the actual amount of water losses include topography, network length, number of branches and service standards, maintenance quality, and network performance. In a well-functioning network, water losses must be continuously controlled. A comprehensive analysis of the water distribution network can be effective in evaluating the existing conditions. this method makes possible appropriate decisions in the design of the hydraulic model of Non-Revenue Water (NRW) which is the reason the present work was written. Consumption management activities may change the time and size of new distribution network facilities such as sources and reservoirs, transmission, and treatment facilities. Consumption management can save water and financial resources. The implementation of the water consumption management program leads to a reduction in costs. Water consumption management and demand forecasting are prerequisites for evaluating the cost-benefit of implementing programs and understanding the challenges facing water systems. Forecasting the demand in the distribution networks of an isolated, sparsely populated area is considered one of the basic measures. During the water demand forecasting period, the water use efficiency program or public education plans are implemented to reduce water consumption. Prediction of water consumption or demand is an important part of the planning process in terms of consumption management plans. Predictions range from simple estimates based on population growth to Predictions based on complex models that consider different variables in water consumption, and the change the predictions can be made for the whole network. the predictions for different groups of consumers (residential, commercial, public) are more accurate. Today, there are various methods to collect water consumption data. Approximate estimates are the only method that was used in the past, while with the advancement of engineering tools and equipment, today more accurate measurements can be used to estimate the amount of consumption of each device. Investigating different consumption management options to prevent water loss should include all types of analysis including cost-benefit analysis. The cost-benefit analysis includes current production cost, income level, profitability of project implementation, and technical, economic, and operational comparison of different projects. In the conditions of lack of financial resources, it is necessary to perform a cost-benefit analysis. To prevent water wastage, household consumers can control the uses outside their home, such as gardening, and car washing, through different methods. The first step in the scientific exploitation of the water systems is to update the map information of the facilities in the computer and the form of geographic information system (GIS). Considering the vast amount of operational information, while updating the map information of facilities in GIS format, different data can be extracted from it in minimum time. This work aimed geo-referenced computer model that can be the solution to the problems of water facilities. The paper presents a geospatial information system (GIS) as one of the efficient methods for the control of a set of reservoirs used for water distribution networks in compliance with the internet of things (IoT). Another novelty of this study informs engineers about the state of the use of remote sensing (RS) facilities equipped with networked sensors, advanced modems, data loggers, and the IoT. This can lead to reducing water loss and non-revenue water (NRW) for saving drinking water. The present work also investigated GIS as a high-precision and quick method of incorporation with the RS and IoT for rapid data intercommunication to reduce the water retention time in reservoirs. This work finally showed that GIS can be linked to new techniques including RS and IoT which can be serious subjects for future research in the fields of environmental engineering, mechanical engineering, Electrical engineering, and control engineering in universities and industries.

Coimbra, Portugal

Ana Cristina Faria Ribeiro
A. K. Haghi

Contents

List of Figures

List of Tables

Apparent Water Loss Control Based on the Geographic Information Systems (GIS)

Abstract

The value of water production in the twenty-second century will be many times higher compared to the production of water in the twenty-first century. The reason for the increase in the cost of water production is the inevitable implementation of consumption management programs in the future. In other words, water production will be equivalent to the most expensive production resources in the future. In the future, the option of water additional production will be replaced by the option of consumption management. Examining the effect of consumption management on water demand is an analytical basis that reveals the cost of optimization in the production sector. In order to predict the costs of implementing the consumption management program and related activities, it is necessary to performance in order to evaluate the control of non-physical (apparent) losses in distribution networks based on the Geographic Information Systems (GIS). The aim of the current work was to investigate the amount of apparent water loss in the water distribution network. In this regard, the total amount of apparent losses was calculated for the number of 200 consumers in an isolated area. The results of the work showed that the total amount of apparent losses was 4.86 (m^3/day).

Keywords

Water loss • Isolated area • Meter • Water distribution network • Consumers

© The Author(s), under exclusive license to Springer Nature Switzerland AG 2025
A. C. F. Ribeiro and A. K. Haghi, *Smart Water Resource Management*, Synthesis
Lectures on Emerging Engineering Technologies,
https://doi.org/10.1007/978-3-031-60799-8_1

1.1   Introduction

In the logic of statistical mathematics, the estimation of consumption information is often done based on the information that has already been collected in this area. In this way, based on the previous observations of consumption information and also some indicators, a logical relationship is established between the information and the mentioned indicators. Among the predictive models that have been developed in this regard, we can mention linear and non-linear regression models, artificial neural networks, fuzzy regression model and neuro-fuzzy model. All models are first trained with observational data. Then they are tested and controlled with experimental data. If the output of the model has a slight difference with the observed actual value, the mathematical model is accepted as a predictive model. There are various methods to collect final consumption information. Approximate estimates were the only method that was used in the past, while with the progress of knowledge and engineering equipment, today more accurate measurements can be used to estimate the amount of water consumption. Types of consumption analysis methods include: intelligent filtering of consumption information, intelligent measurement of common water consumption.

Standard current measuring devices are useful devices in current measurement. The information recorded by these devices can be analyzed manually or by using appropriate software. An information recording device can also be connected to these devices. Using this device, it is possible to record the amount of consumption of all consumers. By installing the transmitter on each of the current measuring devices installed on the consumers' meters, the information can be transferred to the central receiver for recording in the computer. These types of devices can be used to record information in real-time and continuously. Today, the current tracking method is accepted as a very efficient and reliable method. This technique has been used in many consumer component analysis projects around the world. The smart information filtering method consists of a consumption information recording device that is able to collect information. The current work aims to show that the accurate recording of the information of current measuring devices in ideal time intervals can be stored in a consumption analysis system under the consumer information bank. In this case, consumption changes can be detected by using consumption analysis software was [1–4].

1.2   Materials and Methods

This work in 2020 investigated apparent losses in water distribution networks. In this regard, attention was paid to performance indicators in order to evaluate the control of apparent losses in distribution networks. The total amount of apparent losses in the isolated area was calculated for the number of 200 subscribers.

1.2.1 Water Losses

Water losses are divided into two main parts:

- Real (physical) losses.
- Apparent losses (non-physical).

1.2.2 Water Balance

Accurate measurement of drinking water is considered as a component of drinking water production, demand management and loss prevention. In determining the amount of water entering the network and determining losses, the most important issue is the accurate measurement of the amount of water entering the network. The measurement of water production should include the parts of supply sources, purification, entering and leaving water from the network and the amounts of water sent to the distribution networks of isolated areas. Measuring water production is of particular importance in water balance calculations. The steps of water balance calculations are as follows [5–7]:

- Step 1: Determine the volume of water entering the network.
- Step 2: Determine the measured expenses with invoices and the unmeasured expenses with invoices. The sum of these two items is the authorized expenditure with a bill.
- Step 3: The amount of non-revenue water is calculated from the difference between the water entering the network and the revenue-generating water.
- Step 4: Calculate the amount of water measured without a bill and unmeasured water without a bill. The sum of these two items is the allowed expenses without invoice.
- Step 5: The sum of the amounts related to the authorized expenses with invoices and the authorized expenses without invoices is the authorized expenses.
- Step 6: The difference between the amount of input to the network and the allowed consumption is the total amount of losses.

1.2.3 The Fluid Working Pressure and the Pressure of the Water Meter

The working pressure of the fluid in the water supply network must be such that it does not exceed the tolerance threshold of the working pressure of the pipe (pipe burst). In the middle of the night, the water flow in the network is negligible and as a result, the pressure drop reaches its lowest value. The fluid working pressure increases in the low elevation points in pits of the city. The risk of pipe bursting reaches its highest value. The minimum pressure in the network should be enough to have the necessary pressure for the consumer in the pipe at the beginning of each branch. This pressure should meet the following needs [8–11]:

- To compensate for the pressure drop in the branch pipe, water inlet valve and one-way valve.
- Compensate the pressure drop in the water meter. It should be noted that the pressure drop in the water meter, unlike other local pressure drops, is usually high and around 2–6 m.
- To compensate for the pressure drop of the plumbing in each floor of the building.
- Compensate the static height caused by the number of floors of the building.

1.3   Results and Discussion

In this work, in order to determine the total amount of non-physical losses and water balance assessment, it was investigated as follows:

1.3.1   Non-Revenue Water (NRW) Due to Apparent Losses

The amount of apparent losses includes the following [12–14]:

1.3.1.1   Operating Error

- Operation error.
- Failure to read on time.
- Failure to repair meters on time.

Non-revenue water due to exploitation error in isolated areas is as follows:

- Non-revenue water due to error in estimation of unread meters.
- Water without income due to broken meters.
- Water without revenue due to broken meters in one period that were not repaired and replaced in the next period.

In order to use the status of the contours, it includes the following:

- Brooked glass.
- With improper installation location.
- With leakage in upstream side of the meter.
- Unreadable.

- There is damage to the meter screen.
- Apparently healthy, open seal.
- Underground.
- Underwater.
- Steaming.

1.3.1.2 Management Cases

Management cases are as follows.

- Sales without meters.
- Water sales report without volume.
- Failure to report the volume of free water.

1.3.1.3 Human Error

Non-revenue water caused by human error, usually through inaccurate readings by meter reading officers, the lack of order in consumer affairs files, the existence of consumers without files or consumers whose information has not been entered into the computer, and errors occur when entering information into the computer.

1.3.1.4 Non-Revenue Water Caused by a Mistake in the Estimation of Unread Meters

Consumers' meters may not be read in one period for one of the following reasons. The water price of such subscribers is calculated as an estimate. A mistake in estimating the performance of this meter has an effect on the amount of Non-revenue water [15–18].

- When visiting, the consumers' meter reader should not be in the consumers' place.
- The identification code assigned to the consumers is wrong.
- Unreadable meter.
- Cases where the meter is not read by the meter reader. The consumers read the meter and informs.

1.3.1.5 Unreadable Meter

In cases where the meter is not read by the meter reader, the consumer will inform about the reading. Due to the possibility of mistakes in the joint reading, the amount of consumption is considered as an estimated consumption. The amount of non-revenue water resulting from the estimated consumption is as follows (1.1):

$$Q_D = \sum N \times Q_{\text{mean}} - \sum \text{QE} \tag{1.1}$$

1.3.1.6 Non-Revenue Water Due to Broken (Stopped) Meters

To determine the average consumption of such consumers, the average consumption of regular consumers is multiplied by a factor of 1.3 to 1.5.

The amount of non-revenue water caused by damaged meters in one period, which in the next period without repair or replacement, is read as a healthy meter, is calculated as follows (1.2):

$$Q_D = \sum N \times Q_{\text{mean}} - \text{KQE} \tag{1.2}$$

1.3.1.7 Non-Revenue Water from Zero Consumption

Non-revenue water from zero consumption is calculated as follows (1.3).

$$Q_D = \sum N \times Q_{\text{mean}} K \tag{1.3}$$

1.3.1.8 Non-Revenue Water from Consumers Who Do not Have Records

The calculation of non-revenue water from consumers who do not have records is as follows (1.4) [19–26]:

$$Q_D = \sum N \times Q_{\text{mean}} K \tag{1.4}$$

1.3.1.9 Non-Revenue Water from Permitted Branches Without Meter

These consumers have a file, but for reasons such as testing the meter and not re-installing it, they do not have a water meter for a long time. Priced water is not calculated for these consumers. In this way, the water losses resulting from this issue are calculated as follows (1.5).

$$Q_D = \sum N \times Q_{\text{mean}} - \text{KQE} \tag{1.5}$$

Table 1.1 Distribution of consumers' consumption based on consumption rate

No.	Consumption situations	Average flow (lit/h)	Percent (%)
1	High expenses	300	50
2	Average expenses	120	30
3	Low average expenses	60	10
4	Low expenses	30	6
5	Very little expenses	15	4

1.3.1.10 Non-Revenue Water Caused by Unauthorized Branches Without Meters

The amount of non-revenue water from unauthorized branches without meters is calculated as follows (1.6):

$$Q_D = \sum N \times Q_{\mathrm{mean}} K \tag{1.6}$$

1.3.2 Calculation of Consumption and Average Correction Coefficient Based on the Main Method of the World Bank

The distribution of the amount of consumption of consumers based on the consumption rate (Table 1.1) shows that the consumption of water in maximum flows includes the largest share of the consumption of a subscriber. Consumption in minimum discharges is often related to leaks, drips and low common consumption.

The correction factor calculated during the testing of the contours, according to the above percentages, has been applied to the average consumption of the joint in order to finally calculate the actual consumption of the joint. According to the above explanations, the actual shared consumption can be calculated from the following relationship (1.7):

$$Q_{\mathrm{ma}} = Q_m(0.5\mathrm{CF}300 + 0.3\mathrm{CF}120$$
$$+ 1\mathrm{CF}60 + 0.6\mathrm{CF}30 + 0.4\mathrm{CF}15) \tag{1.7}$$

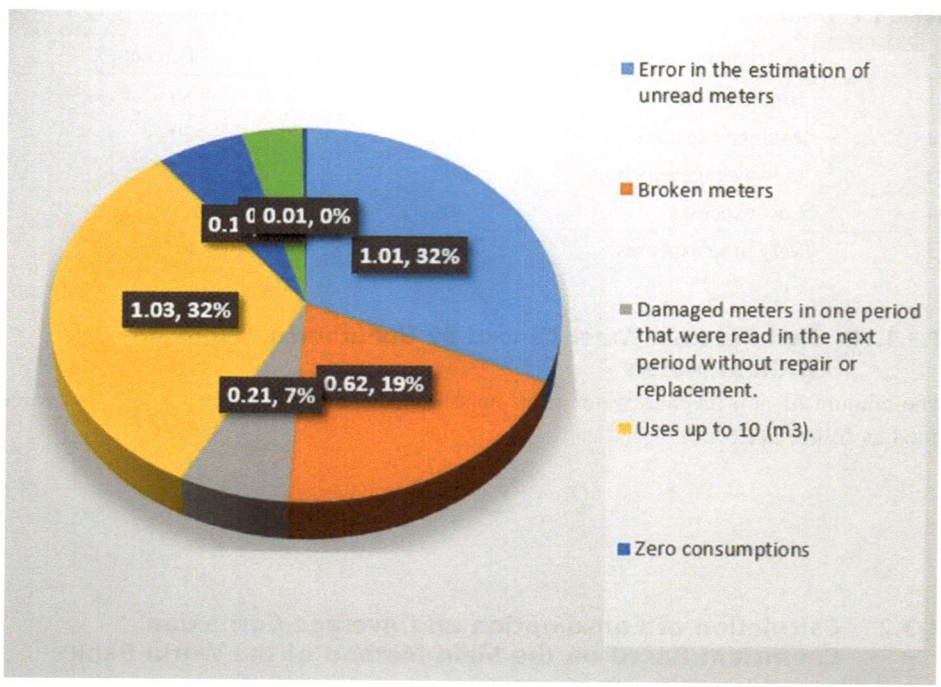

Fig. 1.1 The total amount of non-physical water loss (due to human error) in the isolated area (number of 200 consumers in the isolated area)

In the case that the correction coefficient (CF) becomes infinite in one of the discharges in order to determine the average consumption coefficient (CFm), the relation is corrected. For this case, a compensatory consumption Qad should be considered for the infinite correction coefficient as in Eq. 1.2 be (1.8).

$$Q_{ad} = W_{inf} \times Q_m \tag{1.8}$$

In this formula, Winf is the set of weight coefficients related to the ranges with infinite discharge. That is, the range where the common meter does not show consumption. The total amount of non-physical losses separately in the isolated area for the number of 200 subscribers is as follows (Figs. 1.1 and 1.2, Tables 1.2, 1.3, 1.4, 1.5, 1.6 and 1.7):

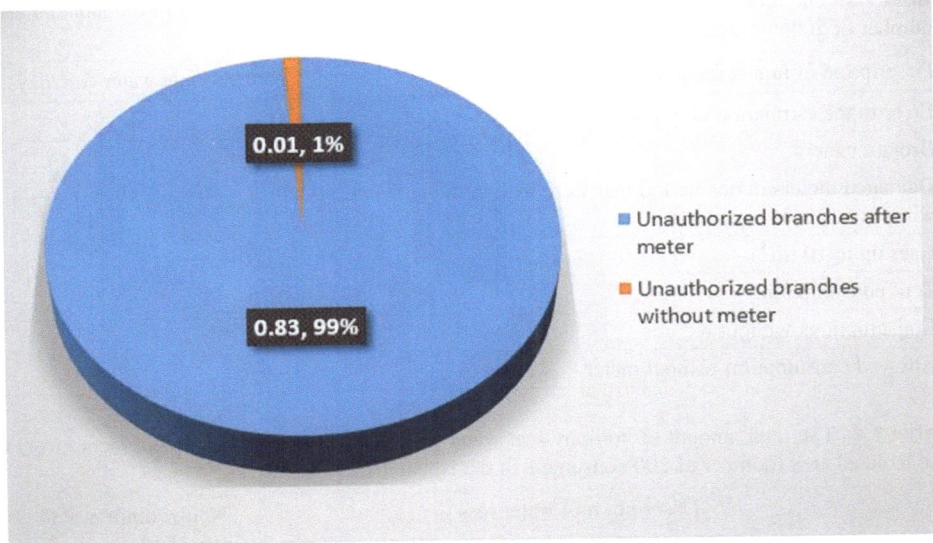

Fig. 1.2 The total amount of non-physical water loss (due to unauthorized uses) in the isolated area (number of 200 consumers in the isolated area)

Table 1.2 The total amount of non-physical water loss (due to human error) in the isolated area (number of 200 consumers in the isolated area)

		Description of non-revenue water	Non-revenue water (m^3/day)
Non physical	Human error	Error in the estimation of unread meters	1.01
		Broken meters	0.62
		Damaged meters in one period that were read in the next period without repair or replacement	0.21
		Uses up to 10 (m^3)	1.03
		Zero consumptions	0.19
		Consumptions without records	0.13
		Allowed consumption without meter	0.01
The sum total of human error			3.2

Table 1.3 The total amount of non-physical water loss (due to human error) in the isolated area (number of 200 consumers in the isolated area)

Description of non-revenue water	Non-revenue water (m³/day)
Error in the estimation of unread meters	1.01
Broken meters	0.62
Damaged meters in one period that were read in the next period without repair or replacement	0.21
Uses up to 10 (m³)	1.03
Zero consumptions	0.19
Consumptions without records	0.13
Allowed consumption without meter	0.01

Table 1.4 The total amount of non-physical water loss (due to instrument measurement error) in the isolated area (number of 200 consumers in the isolated area)

	Description of water loss		Non-revenue water (m³/day)
Instrument measurement error	Consumers' meters	Accuracy error of meter	0
		Manipulation in contour	0.82
	Outflow		0
	The sum total of instrument measurement error		**0.82**

Table 1.5 The total amount of non-physical water loss (due to unauthorized uses) in the isolated area (number of 200 consumers in the isolated area)

	Description of water loss	Non-revenue water (m³/day)
Unauthorized uses	Unauthorized branches downstream of the meter	0.83
	Unauthorized branches without meters	0.01
	The sum of all unauthorized expenses	0.84

Table 1.6 The total amount of non-physical water loss (due to unauthorized uses) in the isolated area (number of 200 consumers in the isolated area)

Description of water loss	Non-revenue water (m³/day)
Branches after meter unauthorized	0.83
Branches without meter unauthorized	0.01

Table 1.7 The total amount of non-physical water loss in the isolated area (number of 200 consumers in the isolated area)

			Description of non-revenue water	Non-revenue water (m^3/day)
Non physical	Human error		Damaged meters in one period were read in the next period without repair or replacement	0.21
			Uses up to 10 (m^3)	1.03
			Zero consumptions	0.19
			Consumptions without records	0.13
			Allowed consumption without meter	0.01
			The sum total of human error	**3.2**
	Instrument measurement error	Consumers' meters	Meter accuracy error	0
			Manipulation in contour	0.82
		Exit from the water facility		0
		The sum total of instrument measurement error		**0.82**
	Unauthorized uses	Unauthorized branches downstream of the meter		0.83
		Unauthorized branches without meters		0.01
		The sum of all unauthorized expenses		**0.84**
	Total non-physical Non-revenue water			**4.86**

1.4 Conclusion

The results of the present work showed that accurate data recording of current measuring devices in 10s intervals contain the appropriate amount of consumption information and desired accuracy. The most scientific method of consumption analysis is the formation of a consumer information bank on the basis of Geographic Information Systems (GIS). Therefore, all water consumption can be stored in a consumption analysis system on the basis of GIS. By using stream consumption analysis software, the relevant consumption changes are detected. The consumption changes are analyzed by the expert, based on the graphic outputs of the software. The final results of the current work in the apparent losses section were calculated as follows:

- Total human error 3.20
- The sum total of tool measurement error is 0.82
- The total amount of unauthorized expenses is 0.84
- The total amount of non-physical water loss is 4.86

The following items are suggested for future works:

- Periodic testing of consumers' meters.
- Reducing leakage in the internal network of consumers.
- Preparing a computer model in the framework of the geographic information system.
- Entering the information collected in the software in the context of the geographic information system.
- Collection of geo-referenced data for model calibration.
- Network calibration through the Internet of Things (IoT).

Abbreviations

Q_D	Non-revenue water due to the error in the estimation of unread meters
N	The number of unread meters
Q_{mean}	Per capita consumption of consumers by dividing consumption classes in terms of cubic meters in one period
Q_E	Estimated consumption amount considered for consumers
Q_D	Amount of non-revenue water due to broken meters
N	Number of defective meters
Q_{mean}	Per capita consumption of subscribers by dividing consumption classes in terms of cubic meters in the third period
Q_E	Estimated consumption amount calculated for consumers with broken meters
K	Coefficient of excess consumption due to lack of commitment to saving
N	Number of consumers with broken meters in the third period and healthy ones read in the fourth period without repairing or replacing the meter
Q_{mean}	Per capita consumption of subscribers divided by usage in cubic meters
Q_E	Amount of consumption calculated for said consumers
Q_D	The amount of non-revenue water caused by broken meters in one period, which were read as healthy meters in the next period without repair or replacement
Q_D	Non-revenue water related to zero consumption

N	Number of subscribers with unrealistic zero consumption
Q_{mean} (m³/day)	Per capita consumption of consumers by usage
K	Coefficient of excess consumption due to lack of commitment to savings (m³)
Q_D	Non-revenue water from the consumption area of consumers without a file
N	Number of consumers without files
Q_{mean}	Per capita consumption of subscribers, divided by usage in terms of m³
K	Coefficient of excess consumption due to lack of commitment to savings
Q_D	The amount of non-revenue water resulting from permitted branches without meters
N	Number of branches allowed without counters
Q_{mean}	Per capita consumption of subscribers divided by usage
K	Coefficient of excess consumption due to lack of commitment to savings
Q_E	Amount of consumption considered for subscribers with authorized branching without meter
UFW	Non-revenue water, unauthorized branches without meters
N	Number of unauthorized branches without meters
Q_{mean}	Per capita consumption of consumers by user division
K	Coefficient of excess consumption due to lack of commitment to savings (In this way, the amount of water without income caused by unauthorized consumption without a meter)
Q_{ma}	Common consumption based on the average correction factor (liters/day)
Q_m	Average common consumption, during the reading period (liters/day)
300CF	Correction coefficient in flow 300 (liters/hour)
120CF	Correction coefficient in flow 120 (liters/hour)
60CF	Correction factor in flow of 60 (liters/hour)
30CF	Correction factor in flow of 30 (liters/hour)
15CF	Correction coefficient in flow 15 (liters/hour)

References

1. Zhang, T., (2006), The Application of GIS and CARE-W on Water Distribution Networks in Skärholmen Pressure Zone, Stockholm, Sweden. Pipeline Technology 2006 Conference.

2. Dong, C., Lei, Z., and Liu, F., (2011), "Internet Quality Abnormal Analysis with k-means Clustering", The Journal of China Universities of Posts and Telecommunications, 18(2), 94–100.
3. Hariri Asli H., and Hozori, A., (2021), "Non-Revenue Water (NRW) and 3d Hierarchical Model for Landslide", Larhyss Journal, 48, 189–210. http://larhyss.net/ojs/index.php/larhyss/article/view/810/810.
4. Papamichail, G.P., and Papamichail, D.P., (2007), "The k-means Range Algorithm for Personalized Data Clustering in E-commerce", European Journal of Operational Research, 177(3), 1400–1408.
5. Haselbach, L., Adesina, M., Muppavarapu, N., et al., (2023), "Spatially Estimating Flooding Depths from Damage Reports", Natural Hazards 117, 1633–1645. https://doi.org/10.1007/s11069-023-05921-2.
6. Weidner, J., Collins, J., Benitez, M., Adesina, M., and Lozoya, C., (2019), Development of a Robust Framework for Assessing Bridge Performance using a Multiple Model Approach, University of Texas at El Paso. Department of Civil Engineering, Report Number: CAIT-UTC-NC39. https://rosap.ntl.bts.gov/view/dot/48948.
7. Asli, H.H., (2023), Applications of Networked Sensors and the Internet of Things (IoT) for Water Treatment. Sustainable Water Treatment and Ecosystem Protection Strategies. Hard ISBN: 9781774915189. https://www.appleacademicpress.com/sustainable-water-treatment-and-ecosystem-protection-strategies-/9781774915189.
8. Du Plessis, J.A., and Hoffman, J.J., (2015), "Domestic Water Meter Accuracy", WTT Transactions on Ecology and the Environment, 200, 197–208.
9. Asli, H.H., and Arabani, M., (2022), "Analysis of Strain and Failure of Asphalt Pavement. Computational Research Progress in Applied Science & Engineering", Transactions of Civil and Environmental Engineering, 8, 1–11. Article ID: 2250. https://doi.org/10.52547/crpase.8.1.2250.
10. Farley, M., and Trow, S., (2007), "Losses in Water Distribution Networks", IWA Publishing (UK).
11. Nayar, V., (2013)., "The Water Crisis—Rethinking Water Governance", Journal of Land and Rural Studies, 1(1), 75–94.
12. Hariri Asli, H., and Nazari, S., (2021), Water Age and Leakage in Reservoirs: Some Computational Aspects and Practical Hints", Larhyss Journal, 48, 151–167. http://larhyss.net/ojs/index.php/larhyss/article/view/808/807.
13. Foster, H.S., and Beattie, B.R., (1981), "On the Specification of Price in Studies of Consumer Demand Under Block Price Scheduling", Land Economics, 57(4), 624–629.
14. Liu, H.-H, and Ong, Ch-Sh, (2008), "Variable Selection in Clustering for Marketing Segmentation Using Genetic Algorithms", Expert Systems with Applications, 34, 52–68.
15. Asli, H.H., (2023), Modeling of Corrosion for Water System by Networked Sensors and the Internet of Things (IoT) in Compliance with Geography Information System (GIS). Sustainable Water Treatment and Ecosystem Protection Strategies. Hard ISBN: 9781774915189. https://www.appleacademicpress.com/sustainable-water-treatment-and-ecosystem-protection-strategies-/9781774915189.
16. Asli, H.H., Arabani, M., and Golpour, Y., (2020), "Reclaimed Asphalt Pavement (RAP) Based on a Geospatial Information System (GIS)", Slovak Journal of Civil Engineering, 28(2), 36–42. https://doi.org/10.2478/sjce-2020-0013. Slovak University of Technology in Bratislava, Slovak.
17. Hariri Asli, K., Hariri Asli, H., Motlaghzadeh, K., and Hariri Asli, K., (2013), "Numerical Techniques in Water Transmission", Frontiers of Engineering Mechanics Research (FEMR), August, 2(3), 56–62, ISSN: 2306–6016 (Online), ISSN: 2306–6024 (Print), published by the world academic publishing co., limited, Hong Kong, Corpus ID: 108917427. http://www.academicpub.org/femr/; https://api.semanticscholar.org/CorpusID:108917427.

18. Hariri Asli, K., and Hariri Asli, K., (2022), Isolated Pressure Zones Based on GIS as a Solution for Water Network Problems. Water Practice and Technology. https://doi.org/10.2166/wpt.2022.119.

19. Foster, S., and Ait-Kadi, M., (2012), "Integrated Water Resources Management (IWRM): How Does Groundwater Fit In?", Hydrogeology Journal, 20(3), 415–418.

20. Wan Mohd, W.M., Beg, A.H., Herawan, H., and Rabbi, K.F., (2012), "An Improved Parameter less Data Clustering Technique based on Maximum Distance of Data and Lioyd k-means Algorithm", Procedia Technology, 1(0), 367–371.

21. Hariri Asli, K., Hariri Asli, K., and Nazari, S., (2023), "Computational Fluid Dynamics Analysis for Smart Control of water supply", Water Supply. https://doi.org/10.2166/ws.2023.306.

22. Berry, M.J., and Linoff, G.S., (2004), Data Mining Techniques: For Marketing, Sales, and Customer Relationship Management. John Wiley and Sons.

23. Hariri Asli, K., and Hariri Asli, K., (2023), Minimum Night Flow (MNF) and Corrosion Control in Compliance with Internet of Things (IoT) for Water Systems, Water Practice and Technology. https://doi.org/10.2166/wpt.2023.012; https://doi.org/10.2166/wpt.2023.012/93513/Minimum-night-flow-MNF-and-corrosion-control-in.

24. Hariri Asli, K., and Hariri Asli, K., (2023), "Smart Water System and Internet of Things", Journal of Modern Industry and Manufacturing, 2(5). https://doi.org/10.53964/jmim.2023005; https://www.innovationforever.com/article.jmim20230111.

25. Nanopoulos, A., Papadopoulos, A.N., and Manolopoulos, Y., (2007), "Mining Association Rules in Very Large Clustered Domains", Information Systems, 32, 649–669.

26. Hariri Asli, K., and Hariri Asli, K., (2023), "Smart Heating, Ventilating, Air-conditioning and Refrigeration by Web-based Geographic Information System", Journal of Modern Industry and Manufacturing, 2(6). https://doi.org/10.53964/jmim.2023006; https://www.innovationforever.com/article.jmim20230139.

Computational Models for Physical Non-revenue Water (NRW) Through Internet of Things (IoT)

2

Abstract

The world today pays serious attention to water reliability, economic efficiency and sustainable development. The topic of loss control in water facilities is of particular importance. The effective factors in the actual amount of water losses include topography, network length, number of branches and service standards, maintenance quality and network performance. In a well-functioning network, water losses must be continuously controlled. In this work, performance indicators were discussed in order to evaluate the control of physical losses in distribution networks. In order to determine the average water pressure in different points of the distribution network, the results of manometer and calibrated hydraulic model were used. For the isolated area, the average pressure during the day and night was taken into account. Based on local measurements, the percentage of reduction of pressure inside the pipe after the fracture was estimated to be about 50% compared to the normal pressure. Continuity equation and Bernoulli's equation were used to determine the amount of Non-Revenue Water due to broken pipes. The amount of physical Non-Revenue Water was calculated in an isolated area with 200 consumers' cases. As a result of the work, the total amount of physical losses was 1.92 (m³/day).

Keywords

Water loss • Working pressure • Pressure curves • Water distribution network • Calibrated hydraulic model

2.1   Introduction

Water consumption management is not possible without water loss control. Physical loss as a part of water loss is an important indicator in evaluating the efficiency of water distribution networks. The increase in physical losses of water indicates a weakness in the design and construction of the network and due to its unscientific maintenance, from the financial and environmental point of view, it is better to express physical losses as a percentage of the volume of water entering the network. By classifying the components of physical losses in the following way, a better understanding of the subject of water losses is obtained:

• Background losses in the form of undetectable small leaks are usually associated with low flow rate, long loss period, and high volume.
• Losses caused by leaks and bursts reported to the network and are usually characterized by high flow rates, short duration, and medium volumes.
• Losses resulting from unreported bursts to the network, detected by active loss control with an average flow rate and volume that depends on the quality of the active leak detection program.
• Overflow or leakage in tanks.

Researches published in England and Japan show that by applying pressure changes in the network, the overall leakage rate increases. This increase in leakage is far beyond the square ratio between pressure and velocity, generally the reason for this phenomenon is that the effective surface of many leaks changes with pressure. For large networks, assuming a linear relationship between pressure and leakage rate is an acceptable simplification [1–4].

There are specific methods for evaluating physical losses beyond the method presented for calculations related to water balance, which are:

• Night flow analysis based on the data obtained from the zoning of the isolated area.
• Recording the number and type of leaks and fractures and their average flow rate and duration.
• Leakage modeling in such a way that it is possible to evaluate the relationship between real network losses and pressure.

The presented definition does not include physical losses after the location of the consumers' meter, in some cases the losses in the consumers' sector are significant and are considered in the discussions of demand management.

The best choice for the most effective performance indicator in the field of physical losses is to express the length of the main lines of the network in terms of kilometers or the number of branches related to it. International experiences show that the most water loss occur on branches. In areas where the density of branches is low, the losses on the main lines of the network are higher.

The aim of the present work is to check the techniques of calculating physical losses in distribution networks in areas that have a higher density of branches [1, 5–7].

2.2 Materials and Methods

This work in 2020 investigated physical losses in distribution networks. In this regard, attention was paid to performance indicators in order to evaluate the control of physical losses in distribution networks. The total amount of physical and non-physical losses in the isolated area was calculated for the number of 200 consumers. Providing any kind of quantitative and qualitative analysis requires more accurate information about the occurrence of incidents. This work carried out the research process during a period of one month:

- Determining the appropriate strategy in dealing with incidents.
- Evaluating the efficiency of different incident response teams.
- Evaluation of the method adopted to perform system maintenance and repair operations.
- Calculation of water losses.
- Calculation and estimation of costs caused by accidents.
- Planning to correct matters and rebuild sensitive points of the network.
- Modifying and updating water distribution network maps.
- Determining the weaknesses of the water distribution system.
- Investigating the causes of accidents.

2.2.1 Water Losses

Water losses are divided into two main parts:

- Real (physical) losses.
- Apparent losses (non-physical).

2.2.2 Water Balance

Accurate measurement of drinking water is considered as a component of drinking water production, demand management and loss determination. In determining the amount of water entering the network and determining losses, the most important issue is the accurate measurement of the amount of water entering the network. The measurement of water production should include the parts of supply sources, purification, entering and leaving water from the network and the amounts of water sent to the distribution networks of isolated areas. Measuring water production is of particular importance in water balance calculations. The steps of water balance calculations are as follows [8–11]:

- Step 1: The volume of water entering the network is determined.
- Step 2: Measured expenses with invoices and unmeasured expenses with invoices are specified. The sum of these two items is the authorized expenditure with a bill.
- Step 3: The volume of non-revenue water is calculated from the difference between the water entering the network and the revenue-generating water.
- Step 4: The amounts of water measured without a bill and unmeasured water without a bill are calculated. The sum of these two items is the allowed expenses without invoice.
- Step 5: The sum of the amounts related to the authorized expenses with invoices and the authorized expenses without invoices is the authorized expenses.
- Step 6: The difference between the amount of input to the network and the allowed consumption is the total amount of losses.

2.2.3 Water Losses in the Distribution Network

Non-revenue water in the water distribution network includes the following:

- Water losses due to breakages and incidents (visible leakage).
- Water losses due to small leaks (invisible leaks).
- Water loss from faucets and connections.
- Removal of fire extinguishers.
- Washing the network through the available drain valves.

2.2.4 Water Losses as a Result of Washing the Network Through the Drain Valves

Network drain valves are used for washing. In order to determine the amount of water discharged from the drain valves, Bernoulli's Eq. (2.1) can be considered for the upstream and downstream of the valve [12]:

$$z_1^2 + \frac{V_1^2}{2g} + \frac{P_1}{\gamma} = z_1^2 + \frac{V_2^2}{2g} + \frac{P_2}{\gamma} \quad \text{(Bernoli equation)} \tag{2.1}$$

Due to the lack of height difference between two points upstream and downstream of the valve, z_1 and z_2 are removed from the equation. Due to the exit of water from the pipe, p_2 also reaches atmospheric pressure and is removed from the equation. To determine the speed of water exiting the pipe, considering that the network washing operation is done at night, the average pressure in the network and the average speed of water in the network pipes are considered.

2.2.5 Water Losses Due to Visible Leakage

Non-Revenue Water due to network accidents and branches is one of the visible leaks. The occurrence of breakdowns and accidents is generally due to non-observance of design and implementation standards during the creation of distribution networks. Failures and accidents occur due to urban development and network development programs. Bernoulli relationships (2.2):

$$gz + \frac{V^2}{2} + \frac{P}{\rho} = \text{Cte} \tag{2.2}$$

and continuity (2.3) are used to determine the amount of Non-Revenue Water caused by broken pipes in the distribution network [13]:

$$Q = A \times V \text{ (Continuity equation)} \tag{2.3}$$

2.2.6 Water Losses Due to Network Breaks and Branches

Breakages usually happen at night with an increase in pressure in the network. The effective factors in the calculation of water losses related to accidents are as follows:

- Information about distribution network events.
- The average amount of broken pipes throughout the year.
- Calibrated network analysis results using pressure measurement results.

In this work, continuity equation and Bernoulli's equation were used to determine the amount of Non-Revenue Water due to broken pipes (2.4–2.5).

$$gz + \frac{V^2}{2} + \frac{P}{\rho} = \text{Cte} \tag{2.4}$$

$$Q = A \times V, \text{ (Continuity equation)} \tag{2.5}$$

The continuity equation and Bernoulli's equation were written for the upstream and downstream points of the fracture. Since the right side of both relations is equal to a constant value, by making the left side of the written equations equal and replacing the information collected from the location of the pipe break, some parameters were removed from the sides of the equation. Integration of continuity equation and Bernoulli equation in terms of V_2 as below (2.6) is obtained [14–16]:

$$gz_1 + \frac{V_1^2}{2} + \frac{P_1}{\rho} = gz_2 + \frac{V_2^2}{2} + \frac{P_2}{\rho} \tag{2.6}$$

Due to the lack of height difference between water break levels, (points 1 and 2) gz_1 and gz_2 are removed from the sides of the equation and the equation is obtained in terms of V_2 (2.7–2.9):

$$\frac{V_2^2}{2} = \frac{V_1^2}{2} + \left(\left[\frac{P_1}{\rho} \right] - \left[\frac{P_2}{\rho} \right] \right) \tag{2.7}$$

$$\frac{V_2^2}{2} = \frac{V_1^2}{2} + \frac{(P_1 - P_2)}{\rho} \tag{2.8}$$

$$P_1 - P_2 = \Delta P \tag{2.9}$$

The speed of water when leaving the fracture site is obtained from the following Eq. (2.10):

$$V_2 = \left[\frac{(V_1^2 + 2\Delta P)}{\rho} \right]^{\frac{1}{2}} \tag{2.10}$$

2.2.7 The Amount of Allowed Leakage in Water Distribution Network

Using the following relationship, the amount of permissible leakage from pipes and connections has been determined (2.11):

$$L = \frac{(N \times D)\sqrt{P}}{3.3} \tag{2.11}$$

By using the above relationship and determining the number of network inlet connections based on the length of the pipe unit, the amount of leakage allowed can be calculated based on different diameters [17–23].

2.3   Results and Discussion

In this work, the amount of water coming out of the leak (water loss) was calculated by using the continuity relationship and knowing the velocity and cross-sectional area of the water leak. The process of calculating physical losses is as follows.

2.3.1   Fracture Cross Section

In order to determine the cross-sectional area of broken pipelines, the cross-sectional area of broken or pores created due to corrosion was considered as 5% of the total cross-sectional area of each pipe.

2.3.2   Average Speed of Water in the Pipe

Using the results of the calibrated network analysis, for the isolated area, first the flow velocity in each pipe diameter was calculated and then the average velocity in different diameters was determined. This amount was close to each other in different diameters and was considered the same for all diameters.

2.3.3   Analysis of Fluid Pressure During Water Leakage

In order to determine the average water pressure in different points of the distribution network, the results of pressure gauge (Fig. 2.1) and calibrated hydraulic model were used. For the isolated area, the average pressure during the day and night was taken into account.

During water leaks from the pipe, the flow of water through the pipe increases. As a result, the pressure inside the pipe decreases. In this work, based on local measurements,

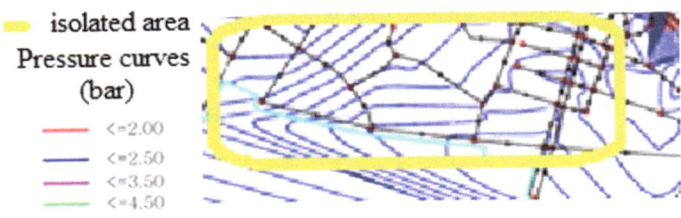

Fig. 2.1   Pressure curves in the isolated area

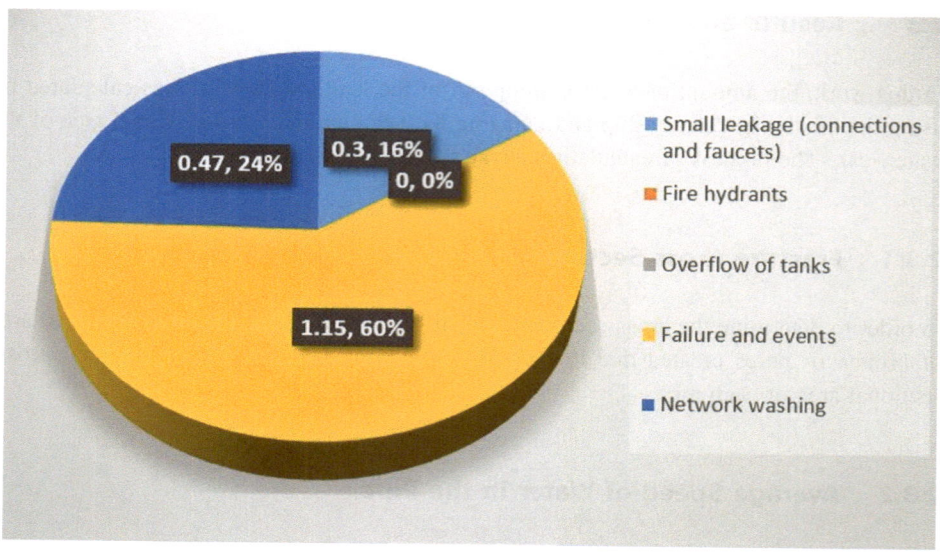

Fig. 2.2 The total amount of physical water loss in the isolated area (number of 200 consumers in the isolated area)

Table 2.1 The total amount of physical water loss in the isolated area (number of 200 consumers in the isolated area)

Description of water loss	Non-revenue water (m^3/day)
Small leakage (connections and faucets)	0.3
Fire hydrants	0
Overflow of tanks	0
Failure and events	1.15
Network washing	0.47

the percentage of the decrease in the pressure inside the pipe after the fracture compared to the pressure in the normal state was estimated to be about 50%.

2.3.4 Calculation of Non-Revenue Water Amounts and Production Estimation

Non-Revenue Water caused by physical losses in the isolated area was calculated (Fig. 2.2, Tables 2.1 and 2.2).

Table 2.2 The total amount of physical water loss in the isolated area (number of 200 subscribers in the isolated area)

Description of Non-revenue water		Non-revenue water (m³/day)
Physical	Small leakage (connections and faucets)	0.3
	Fire hydrants	0
	Overflow of tanks	0
	Failure and events	1.15
	Network washing	0.47
	Total physical Non-revenue water	1.92

2.4   Conclusion

The results of the present work were obtained during field operations and using the results of pressure measurement, modeling, model calibration and network analysis through Internet of Things (IoT). Calculations related to losses caused by accidents and fractures were done with the theoretical method of AWWA. According to the statistics related to the accidents and the review and evaluation of this information, the average amount of these accidents according to the probability of their occurrence per month was included in the calculations. The information used in these calculations was based on information from the distribution network events worksheet, the average amount of pipe breakage, the results of the calibrated network analysis using pressure measurement results. The result of calculating the amounts of non-revenue water and estimating physical losses in the isolated area in this work was 1.92 (m³/day). The following items are suggested for future works:

- Preparing a conceptual model in the framework of the geographic information system.
- Entering the geo-spatial data in the context of the geographic information system.
- Definition of geo-referenced data for model calibration.
- Network calibration through the Internet of Things (IoT).

Abbreviations

g	Acceleration of gravity (m/s^2)
z	Vertical height (m)
v	Fluid speed (m/s)
p	Fluid pressure (pa)
ρ	Density (kg/m^3)
Q	Volume of water past the known cross section (m^3/s)

A	Fracture cross section (m^2)
V_2	Water speed when leaving the fracture section (m/s)
$P_2 = P_0$ (pa)	Pressure outside the pipeline (atmospheric pressure) (pa)
P_1	Pressure inside the pipeline (pa)
V_1	Water speed inside the pipe (m/s)
V_2	Water velocity while leaving the fracture site (m/s)
L	Allowed leakage rate (cm^3/h)
N	Number of connections in the pipeline
D	Pipe diameter (cm)
P	Average pressure inside the pipe (kg/cm^2)

References

1. Mariolakos, I., (2007), "Water Resources Management in the Framework of Sustainable Development", Desalination, 213, 147–151.
2. Asanoa, T., and Cotruvob, J.A., (2004), "Groundwater recharge with reclaimed municipal wastewater: health and regulatory considerations", Journal of Water Research, 38, 1941–1951.
3. Arregui, F.J., Gavara, F.J., Soriano, J., and Pastor Jabaloyes, L., (2018), "Performance Analysis of Ageing Single-Jet Water Meters for Measuring Residential Water Consumption", Water, 10(612), 2–18.
4. Bau, F., Gomes, P., Baran, P., Drouineau, H., Larinier, M., Alric, A., Travade, F., and De Oliveira, E., (2013), Suivi par radiopistage de la dévalaison de l'anguille argentée sur le Gave de Pau au niveau des ouvrages hydroélectriques d'Artix, Biron, Castetarbe, Baigts et Puyoo.
5. Hariri Asli H., and Hozori, A., (2021), Non-Revenue Water (NRW) and 3d Hierarchical Model for Landslide. Larhyss Journal, 48, 189–210. http://larhyss.net/ojs/index.php/larhyss/article/view/810/810
6. Haselbach, L., Adesina, M., and Muppavarapu, N. et al., (2023), "Spatially Estimating Flooding Depths from Damage Reports", Natural Hazards, 117, 1633–1645. https://doi.org/10.1007/s11069-023-05921-2.
7. Weidner, J., Collins, J., Benitez, M., Adesina, M., and Lozoya, C., (2019), Development of a Robust Framework for Assessing Bridge Performance Using a Multiple Model Approach, University of Texas at El Paso. Department of Civil Engineering, Report Number: CAIT-UTC-NC39. https://rosap.ntl.bts.gov/view/dot/48948.
8. Hariri Asli, H., and Nazari, S., (2021), "Water Age and Leakage in Reservoirs: Some Computational Aspects and Practical Hints", Larhyss Journal, 48, 151–167. http://larhyss.net/ojs/index.php/larhyss/article/view/808/807.
9. Asli, H.H., (2023), Applications of Networked Sensors and the Internet of Things (IoT) for Water Treatment. Sustainable Water Treatment and Ecosystem Protection Strategies. Hard ISBN: 9781774915189. https://www.appleacademicpress.com/sustainable-water-treatment-and-ecosystem-protection-strategies-/9781774915189
10. Thornton, J., Sturm, R., Kunkel, G., (2008), "Water Loss Control Manuel". McGraw-Hill, Second Edition. Water Loss Control Committee (2006), "Water Audits and Loss Control Programs", AWWA M36.

11. Asli, H.H., and Arabani, M., (2022), "Analysis of Strain and Failure of Asphalt Pavement. Computational Research Progress in Applied Science & Engineering", Transactions of Civil and Environmental Engineering, 8, 1–11. Article ID: 2250. https://doi.org/10.52547/crpase.8.1.2250.

12. Asli, H.H., (2023), Modeling of Corrosion for Water System by Networked Sensors and the Internet of Things (IoT) in Compliance with Geography Information System (GIS). Sustainable Water Treatment and Ecosystem Protection Strategies. Hard ISBN: 9781774915189. https://www.appleacademicpress.com/sustainable-water-treatment-and-ecosystem-protection-strategies-/9781774915189.

13. Asli, H.H., Arabani, M., and Golpour, Y., (2020), Reclaimed Asphalt Pavement (RAP) Based on a Geospatial Information System (GIS). Slovak Journal of Civil Engineering, 28(2), 36–42. https://doi.org/10.2478/sjce-2020-0013. Slovak University of Technology in Bratislava, Slovak.

14. Moahloli, A., Marnewick, A., and Pretorius, J.H.C., (2019), "Domestic Water Meter Optimal Replacement Period to Minimize Water Revenue Loss", Water SA, 45, 165–173.

15. Hariri Asli K., Hariri Asli H., Motlaghzadeh, K., Hariri Asli K., (2013), "Numerical Techniques in Water Transmission", Frontiers of Engineering Mechanics Research (FEMR), August, 2(3), 56–62, ISSN: 2306–6016 (Online), ISSN: 2306–6024 (Print), published by the world academic publishing co., limited, Hong Kong, Corpus ID: 108917427. http://www.academicpub.org/femr/; https://api.semanticscholar.org/CorpusID:108917427.

16. Hariri Asli K., and Hariri Asli K., (2022), Isolated Pressure Zones Based on GIS as a Solution for Water Network Problems. Water Practice and Technology. https://doi.org/10.2166/wpt.2022.119.

17. 18. Ncube, M., and Taigbenu, A.E., (2019), "Assessment of Apparent Loss Due to Meter Inaccuracy Using an Alternative, Validated Methodology", Water Supply, 19, 1212–1220.

18. Hariri Asli K., Hariri Asli K., and Nazari, S., (2023), Computational Fluid Dynamics Analysis for Smart Control of Water sSupply. Water Supply. https://doi.org/10.2166/ws.2023.306.

19. Winter, H.V., Jansen, H.M., and Bruijs, M.C.M., (2006), Assessing the Impact of Hydropower and Fisheries on Downstream Migrating Silver Eel, Anguilla Anguilla , by telemetry in the River Meuse. Ecology of Freshwater Fish, 15, 221–228.

20. Hariri Asli K., and Hariri Asli K., (2023) Minimum Night Flow (MNF) and Corrosion Control in Compliance with Internet of Things (IoT) for Water Systems. Water Practice and Technology. https://doi.org/10.2166/wpt.2023.012.

21. Asli, KiH., and Asli, KaH., (2023), "Smart Water System and Internet of Things." Journal of Modern Industry and Manufacturing, 2, 5. https://doi.org/10.53964/jmim.2023005; https://www.innovationforever.com/article.jmim20230111

22. Pasha Zanousi, S., Ayati, B., and Ganjidoust, H., (2010), "Investigation of Tubifex Worm's Potential in Mass and Volume Reduction of Sludge Wastewater Treatment Plants in Laboratory Scale", Journal of Water and Wastewater, 24(4), 59–65.

23. Asli, KiH., and Asli, KaH., (2023), Smart Heating, Ventilating, Air-conditioning and Refrigeration by Web-based Geographic Information System. Journal of Modern Industry and Manufacturing, 2, 6. https://doi.org/10.53964/jmim.2023006; https://www.innovationforever.com/article.jmim20230139.

Smart Management of Water Loss at the District Metered Areas (DMA)

3

Abstract

A comprehensive analysis of the water distribution network can be effective in evaluating the existing conditions. this method makes possible appropriate decisions in the design of the hydraulic model of Non-Revenue Water (NRW). It also informs the calculation of physical losses and the implementation of consumption management. Access to the data related to the maintenance of the water system can be useful in the correct analysis of the conditions. The data of water flow entering the District Metered Areas or DMA's in the water distribution network can be measured by regional meters. Choosing the place to install regional meters requires network analysis to determine the range of flow and pressure changes. In this work, along with pressure measurement and calibrating the hydraulic model, a computer map of pressure curves was prepared. After creating the isolation area, installing the meter in DMA showed that physical losses are under the direct influence of fluid working pressure changes. In the regression analysis, the fluid working pressure in the water distribution network as a dependent variable of changes in the outflow water of the reservoir (incoming water flow to the DMA) had a P-value of 0.989.

Keywords

Fluid working pressure • Hydraulic model • Regression analysis • Isolated area • Water losses

© The Author(s), under exclusive license to Springer Nature Switzerland AG 2025
A. C. F. Ribeiro and A. K. Haghi, *Smart Water Resource Management*, Synthesis
Lectures on Emerging Engineering Technologies,
https://doi.org/10.1007/978-3-031-60799-8_3

3.1   Introduction

Regional meters are a tool for evaluating the effect of the water distribution separation to isolated regions on the hydraulic status of the water network. The hydraulic condition of the regions before and after the isolation of the water distribution network is simulated and compares by a computer model. In this case, the changes created in the model represent the real state of the system after the isolated regions. Examining the information output from the computer model before and after the installation of regional meters (maximum and minimum water consumption) shows the advantages and disadvantages of the proposed options.

Regional meters should be designed to be able to measure fluid flow from the main pipeline of a large size to the main pipeline of a smaller region. Choosing the installation location of regional regions in order to determine the consumption of the region requires an accurate analysis of the hydraulic model. This is to determine the range of fluid flow intensity at different points and to identify the optimal points for the installation of the meter. In this context, the use of the following items can be beneficial [1–3]:

- Regional meters, preferred when the diameter of the pipe is less than 500 mm. It should be selected from the mechanical vane type. Because these types of meters do not need a power source and other accessories, and only a basin should be built to install the meters.
- The characteristics of the region should be considered in choosing the type and size of the meter. For example, in regions that do not have a storage source, the installed meters must have sufficient sensitivity to measure the intensity of Minimum Night Flow (MNF).
- Always move in one direction at the installation location of the fluid flow meter.
- Regional meters should be installed in the places where the maximum intensity of the flow is likely. This will make the measurements have good accuracy.
- If the running water in the pipe is free of foreign and suspended substances, floating fibers and air, the use of rotating vane meters equipped with a filter before the meter is satisfactory. If it is not possible to use the rotary vane meter for reasons such as inappropriate water quality, repair and maintenance problems, the use of electromagnetic meters should be considered.
- In order to allow access to the meters and ease its repair and maintenance, regional meters are usually placed on the by-pass route of the pipeline and by installing non-return valves before and after the meter.
- Electromagnetic meters should be used in places that are not affected by electric power sources and magnetic fields, otherwise they should be completely protected.
- If the minimum flow through the pipes is low, we have to use a meter with a small diameter. it causes the large loss by the use of the meter, the changes and related

connections. it is recommended to use combined meters include ultrasonic and induction meters. Combined meters include a rotary vane meter and a one-way valve and a bypass line with a one-way valve and a meter with a smaller diameter. In the intensity of low currents, the valve of the main meter remains closed and all the flow passes through the small meter, and therefore the flow rate is measured accurately. As soon as the current intensity reaches the predetermined range, the main valve of the meter is opened due to the pressure difference and thus the current intensity is measured from both meters.

- Calibration and periodic testing of meters should be easily possible. For this reason, regional meters can be tested by installing a portable ultrasonic meter or by using a reservoir in the workshop.
- Usually, regional meter readings are done at a certain time of the day. But the real value of these meters is shown when the resulting information is presented to the management of water supply and distribution on a daily and even hourly basis. Therefore, if the regional meters have the ability to provide information in the desired forms and in desired periods, they will help a lot to control water loss [4–8].

3.1.1 Modeling and Hydraulic Analysis

The basis of the hydraulic analysis is the investigation of the water pressure drop in the distribution network pipes. In the hydraulic analysis of the network, the pressure drop in the water network pipes is as follows (3.1):

$$H_L = a \cdot q^b \tag{3.1}$$

The pressure drop in the pipes can be calculated from the three formulas of Hazen-Williams, Darcy-Weisbach and Chezy-Manning. The Darcy-Weisbach formula is often used for laminar flows in pipes. Chezy-Manning formula is also often used for flow in open channels. But the Hazen-Williams formula is widely used in water distribution networks. Table 3.1 shows the coefficients for each pressure drop formula [9, 10].

Table 3.1 Resistance coefficient for formulas of Hazen-Williams; Darcy-Weisbach; Chezy—Manning

Formula	Resistance coefficient	Power
Hazen-Williams	$1.438656c^{-1.85} d^{4.871}$	1.85
Darcy-Weisbach	$0.0077f\ d^{-5}$	2
Chezy-Manning	$1.42\ n^2\ d^{-5.331}$	2

Table 3.2 Roughness coefficient for new pipes

Material	Hazen Williams (c)	Darcy-Wiesbach ε (mm)	Manning (n)
Ductile iron	130–140	0.26	0.012–0.015
Concrete or pre-stressed	120–140	0.3–3.05	0.012–0.017
Galvanized iron	120	0.152	0.015–0.017
Plastic	140–150	0.00152	0.011–0.015
Steel	140–150	0.046	0.015–0.017

3.1.2 Deposition and Water Pressure Drop in the Pipe

Pipes change in diameter under the influence of chemical parameters and fluid deposition in the pipe. The coefficients of fluid working pressure drop should be chosen in such a way that the effect of the above parameters is considered in it. Manufacturers declare the roughness coefficients for the Hazen-Williams formula, the Darcy-Weisbach formula, and the Chezy-Manning formula for new pipes (Table 3.2). These roughness coefficients for all types of new pipes include ductile iron, pre-stressed concrete, galvanized iron, plastic, and steel [11–13].

In every water distribution network, accessories such as one-way valves, butterfly valves, air valves, drain valves, three-way valves, knee valves, and switches are installed. This equipment is used as controllers of the transmission line, and network based on the conditions of the land and plan. With the use of accessories, there will be a pressure drop in the network. Calculations related to such partial pressure drops are accompanied by errors due to complexity. In network analysis, the equivalent length method is used to calculate local losses. Considering the extent of the network and the number of equipment, an additional percentage can be considered for the length of the pipe to allow for local drops. This value is between 5 and 10% for water supply networks. Using a calibrated mathematical model can reduce the complexity of pressure drop calculations. Otherwise, the work efficiency of the water system will decrease. In such a situation, the cost of repairs and maintenance of the water system will increase greatly [14].

3.1.3 Sedimentation and Corrosion in Pipes

Sedimentation causes increased corrosion in pipes and is one of the reasons for increasing water losses in water transmission systems. Based on these reasons, this work with the aim of cost saving due to water loss as one of the most important challenges, studied the physical Non-Revenue Water (NRW) modeling for water systems.

3.2   Materials and Methods

To define the hydraulic model of non-revenue water, first computer maps and information on facilities are evaluated.

3.2.1   Hydraulic Model and Computer Map

To define the hydraulic model, maps of supply, transmission, and storage facilities, as well as distribution network maps, are updated during local survey operations. In this operation, new pipelines along with all new valves, pressure gauges, and other equipment are reflected on these maps. The existing status maps are scanned and edited. During the process of scanning and editing the maps, the technical specifications of network facilities and equipment are updated in the form of DMA or isolated areas in compliance with the Geographic Information System (GIS) (Fig. 3.1). The actions after scanning and editing the maps are as follows [15–18]:

- Integrating and replicating maps.
- Alignment of maps in terms of font, title, and standard signs of facilities including pipes, faucets, pressure gauges, and switches.
- Updating the computer maps of the water distribution network.

3.2.2   Hydraulic Model and Pressure Gauge

Pressure gauges are read during the schedule. The results of pressure measurement are checked with the chart of pressure changes at different hours of the day and night (Fig. 3.2). The pressure measurement results are formed into pressure curves for District Metering Area (DMA).

3.2.3   Hydraulic Model and Pipes

The structure of the distribution network consists of small-diameter and large-diameter pipes. Pipes are only part of the water distribution network, which is always buried underground. Pipes are more vulnerable than other components of the network. Pipes are damaged due to dynamic loads. Maintaining the network does not only mean visiting when accidents occur. A regular visit schedule is required, considering the priority of the old network areas. Before causing any costly damage, it is necessary to use the network leak detector during visits. Usually, the amount of water lost from small leaks in

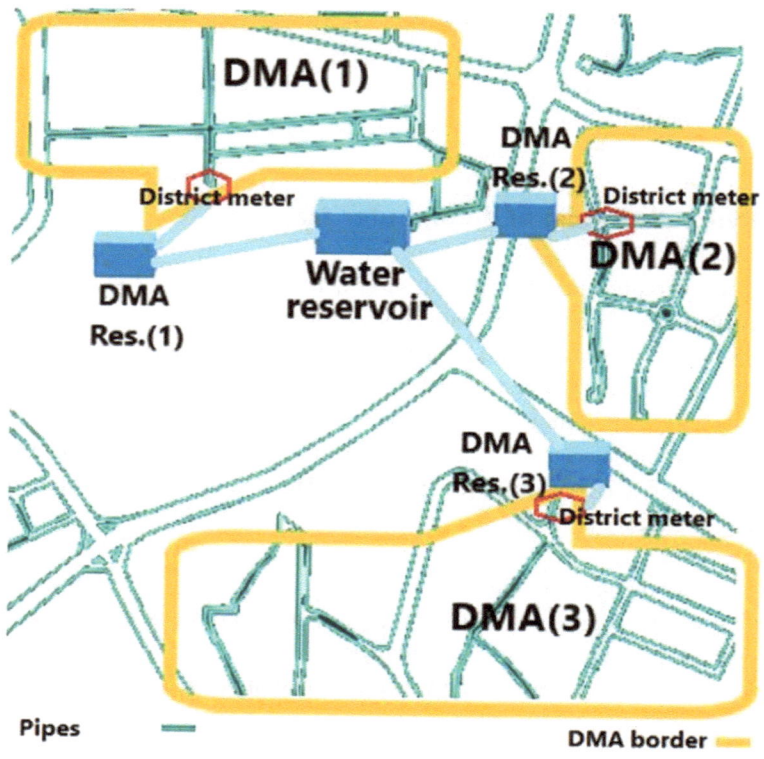

Fig. 3.1 Hydraulic model for District Metering Area (DMA)

the network is more than the large leaks that cause accidents. The reasons for reducing
the transfer capacity and water losses in the pipes are as the following [19–24]:

- Entry of external materials into the pipes, such as wood, and stone, which enter the
  pipes during pipe laying or necessary repairs.
- Sludge settling, iron, manganese, and plant growth.
- Creating an inner layer of sediment on the inner wall of the pipes.
- Prominences caused by the oxidation of the metal pipe or the chemical compounds of
  the water.
- Air accumulation in the highest points of the grid and transmission lines.
- Reducing the cross-sectional area of the pipe due to sedimentation of faucets in semi-
  closed states or bending of pipes.

To prevent water losses, it is necessary to carefully implement the network, use high-
quality materials, and carefully monitor its implementation. Compliance with the technical

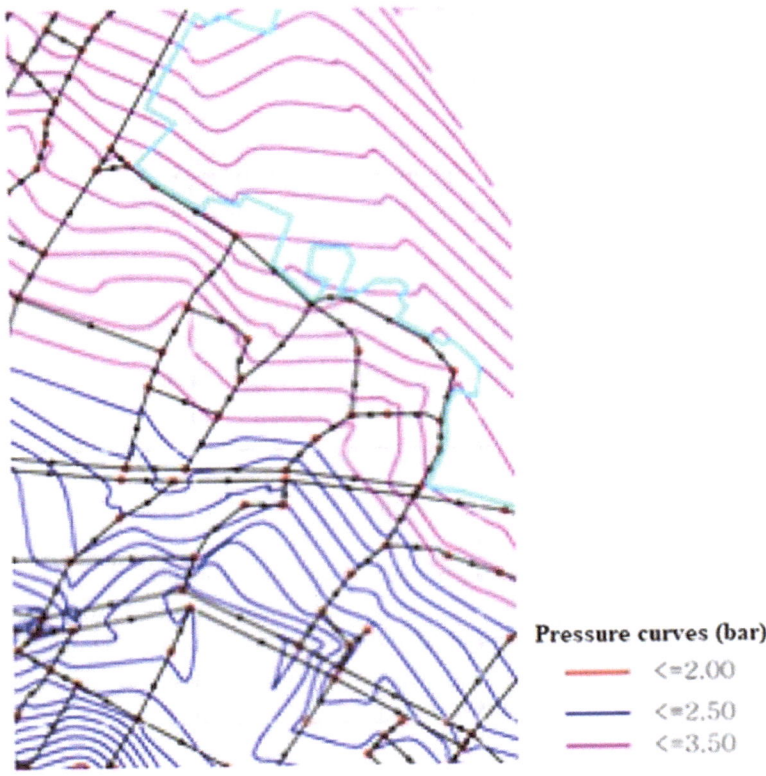

Pressure curves (bar)
——— <=2.00
——— <=2.50
——— <=3.50

Fig. 3.2 Hydraulic model pressure curves for District Metering Area (DMA)

principles of implementation is one of the factors that can have a great effect on the health
of the network and increase its life. In the process of controlling water loss in facilities,
the following measures should be taken:

- Full air evacuation control throughout the length of the pipeline.
- Washing pipes.
- Visiting the accessible parts of the pipe to control the layer of sediment material created
  on the surface of the inner wall of the pipes.

3.2.4   Reservoirs and Regional Meters

Before installing regional meters, the following are checked:

- Tank overflow system is ready for operation.
- The performance of the taps entering the tanks is guaranteed.
- The performance of the outlet taps from the tanks is guaranteed.

3.3   Results and Discussion

At present work, for preparation of the computer model and model calibration was investigated the results of pressure measurement of the water distribution network.

3.3.1   Meters and Pressure Gauges

This work investigated physical losses in water distribution networks. In this regard, attention was paid to performance indicators in order to evaluate the control of physical losses in distribution networks. Identifying the optimal points for installing the meters and pressure gauges was carried out after analyzing the network and pressure measurement. The activity of pressure gauge reading was done for DMA.

3.3.2   Hydraulic Model Calibration by District Meters

During the field operation, an Ultrasonic Flow Meters (UFM) as a district meter was selected for the water distribution system. The hydraulic model calibration process was defined in compliance with WATERGEMS software and by drawing pressure curves. The calibrated model revealed the range of the water loss. This procedure showed the error of the measuring equipment and the correction of these errors. It provided smart management for water loss at the DMA of distribution network.

Water working pressure (bar)

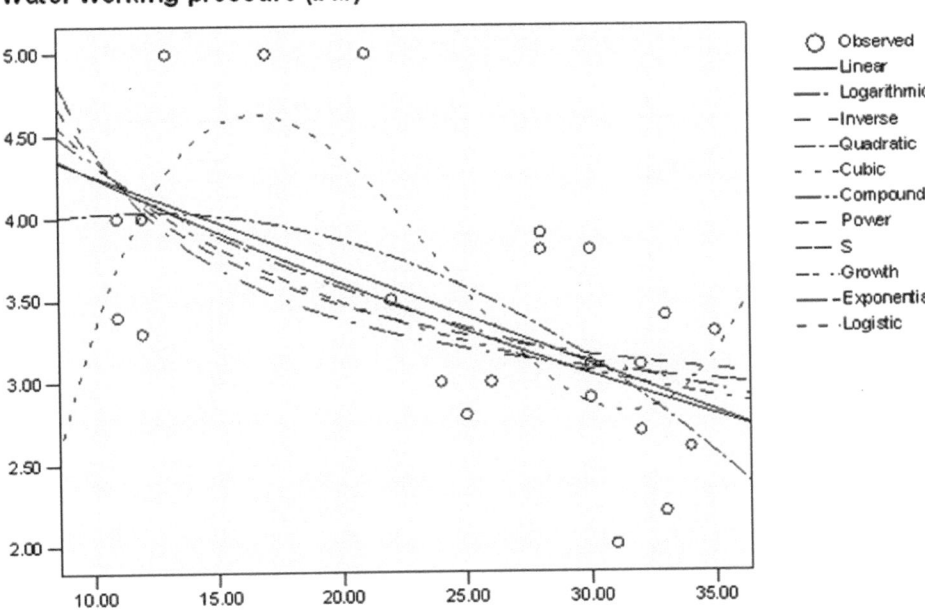

Outflow of reservoir (lit/sec)

Fig. 3.3 Scatter diagram for dependent variable, water working pressure (bar) and independent variable, outflow of the reservoir (lit/sec)

3.3.3   Hydraulic Model and Regression Analysis

The outflow of the reservoir towards DMA was compared with the water working pressure at the distribution network. This procedure led to the clarification and correction of the amount of physical NRW and its components. Regression analysis investigated the dependent variable such as the water working pressure (bar) and independent variables (3.2–3.3) such as the outflow of the reservoir (lit/sec) (Fig. 3.3, Tables 3.3, 3.4 and 3.5).

$$\text{Water working pressure} = f\,(\text{Output of district meter}) \qquad (3.2)$$

(Dependent variable): Water working pressure (bar); (Independent variable): outflow of the reservoir (lit/sec).

$$P = f(Q) \qquad (3.3)$$

Table 3.3 Flow metering by UFM (district meter) for the distribution network

Time of metering (h)	Outflow of reservoir (lit/sec)	Water working pressure (bar)
1	12	3.3
2	11	3.4
3	12	4
4	11	4
5	13	5
6	17	5
7	21	5
8	22	3.5
9	24	3
10	26	3
11	26	3
12	25	2.8
13	32	2.7
14	31	2
15	33	2.2
16	34	2.6
17	30	2.9
18	30	3.1
19	32	3,1
20	33	3.4
21	35	3.3
22	30	3.8
23	28	3.8
24	28	3.9

Table 3.4 Model Summary and Parameter Estimates; Dependent Variable: Water working pressure (bar); The independent variable is Outflow of reservoir (lit/sec)

Equation	Model summary					Parameter estimates			
	R Square	F	df_1	df_2	Sig.	Constant	b_1	b_2	b_3
Linear	0.337	11.158	1	22	0.003	4.839	−0.058		
Logarithmic	0.294	9.151	1	22	0.006	6.874	−1.101		
Inverse	0.240	6.947	1	22	0.015	2.549	18.400		
Quadratic	0.367	6.078	2	21	0.008	3.627	0.068	−0.003	
Cubic	0.495	6.546	3	20	0.003	−6.910	1.692	−0.078	0.001
Compound	0.336	11.127	1	22	0.003	5.034	0.983		
Power	0.296	9.231	1	22	0.006	9.139	−0.322		
S	0.245	7.140	1	22	0.014	0.947	5.414		
Growth	0.336	11.127	1	22	0.003	1.616	−0.017		
Exponential	0.336	11.127	1	22	0.003	5.034	−0.017		
Logistic	0.336	11.127	1	22	0.003	0.199	1.017		

Table 3.5 Test Statistics; Regression analysis for the dependent variable and independent variables

Parameters	Water working pressure (bar)	Outflow of reservoir (lit/sec)	Time of metering (h)
Chi-Square	4.750	4.000	0.000
df	14	15	23
Asymp. Sig.	0.989	0.998	1.000

3.4   Conclusion

In this work, the information on the following items was investigated to the preparation of the hydraulic model for the water distribution system:

- Pipes.
- Faucets.
- Pumps.
- Nodes.
- Water flow fluctuations.
- Water consumption fluctuations.
- Water consumption patterns.
- Water loss coefficient.

The installation location of the regional meters was determined to calculate the water consumption of the region. The analysis of the hydraulic model led to the calculation of the variation interval of fluid flow at different points of the network. This process identifies the optimal points for installing the meter.

The installation of a regional meter in the DMA or isolated area revealed the following facts:

- Physical losses were under the direct influence of fluid working pressure changes.
- In the regression analysis for the hydraulic data of the DMA isolated area, the data related to the working pressure of the fluid in the water distribution network (dependent variable) and the data related to the changes in the water flow coming out of the tank (independent variable) had a significant relationship.
- In the regression analysis, the P-value showed 0.989.

Suggestions for Future Research

The air intake into pipes can cause oxidation of rebar in concrete reinforced pipes and steel pipes in the presence of chlorine and cause water leakage or physical NRW. The water loss lead to economic losses. The following items are suggested to those who are interested in studying the use of a smart NRW model by IoT based on the GIS:

- Define smart NRW model by IoT based on the GIS.
- Installation of a regional meter in the DMA in compliance with web-based GIS.

Abbreviations

C	Roughness coefficient
d	Pipe diameter
f	Friction factor
n	Manning's Roughness coefficient
HL	Pressure drop
q	Flow rate
a	Resistance coefficient
b	Power
P	Water working pressure (bar)
Q	Inflow to the reservoir (lit/sec)

References

1. Wu Zheng, Y., Sage, P., and Turtle, D., (2010), "Pressure-dependent Leak Detection Model and its Application to a District Water System", Journal of Water Resources Planning and Management, 136(1), 116–128.
2. Hariri Asli H., and Hozori, A., (2021), "Non-Revenue Water (NRW) and 3d Hierarchical Model for Landslide", Larhyss Journal, 48, 189–210. http://larhyss.net/ojs/index.php/larhyss/article/view/810/810.
3. Farley, B., Mounce Stephen, R., and Boxall Joby, B., (2011), Field Validation of "Optimal" Instrumentation Methodology for Burst/leak Detection and Location. Water Distribution Systems Analysis Conf. ASCE, Reston, VA
4. Asli, H.H., (2023) Applications of Networked Sensors and the Internet of Things (IoT) for Water Treatment. Sustainable Water Treatment and Ecosystem Protection Strategies. Hard ISBN: 9781774915189. https://www.appleacademicpress.com/sustainable-water-treatment-and-ecosystem-protection-strategies-/9781774915189.
5. Asli, H.H., (2023), Modeling of Corrosion for Water System by Networked Sensors and the Internet of Things (IoT) in Compliance with Geography Information System (GIS). Sustainable Water Treatment and Ecosystem Protection Strategies. Hard ISBN: 9781774915189. https://www.appleacademicpress.com/sustainable-water-treatment-and-ecosystem-protection-strategies-/9781774915189.
6. Goulet, J.-A., Coutu, S., and Smith, I.F.C., (2013), "Model Falsification Diagnosis and Sensor Placement for Leak Detection in Pressurized Pipe Networks", Advanced Engineering Informatics, 27(2), 261–269.
7. Asli, H.H., and Arabani, M. (2022), "Analysis of Strain and Failure of Asphalt Pavement. Computational Research Progress in Applied Science & Engineering", Transactions of Civil and Environmental Engineering, 8, 1–11. Article ID: 2250. https://doi.org/10.52547/crpase.8.1.2250.
8. Ayati, A.H., Haghighi, A., and Lee, P., (2019), "Statistical Review of Major Standpoints in Hydraulic Transient-based Leak Detection", Journal of Hydraulic Structures 5(1), 1–26.
9. Hariri Asli K., and Hariri Asli K., (2022), Isolated Pressure Zones Based on GIS as a Solution for Water Network Problems. Water Practice and Technology. https://doi.org/10.2166/wpt.2022.119
10. Xu, X., and Karney, B., (2017), An Overview of Transient Fault Detection Techniques, in Modeling and Monitoring of Pipelines and Networks: Advanced Tools for Automatic Monitoring and Supervision of Pipelines, C. Verde and L. Torres, Editors. Springer International Publishing, Cham, pp. 13–37.
11. Asli, H.H., Arabani, M., and Golpour, Y., (2020), Reclaimed Asphalt Pavement (RAP) Based on a Geospatial Information System (GIS). Slovak Journal of Civil Engineering, 28(2), 36–42. https://doi.org/10.2478/sjce-2020-0013. Slovak University of Technology in Bratislava, Slovak.
12. Plath, M., Mathias, E., and Knut, W., (2014), "Energy Efficiency and Energy Saving in the German Water Industry", Water Practice & Technology, 9(2), 256–263.
13. Hariri Asli K., Hariri Asli K., and Nazari, S., (2023), Computational Fluid Dynamics Analysis for Smart Control of Water Supply. Water Supply. https://doi.org/10.2166/ws.2023.306.
14. Wang, X.-J., et al., (2002), "Leak Detection in Pipelines Using the Damping of Fluid Transients", Journal of Hydraulic Engineering, 128(7), 697–711.
15. Ferrante, M., Brunone, B., and Meniconi, S., (2009), "Leak-edge Detection", Journal of Hydraulic Research, 47(2), 233–241.

16. Asli, KiH., and Asli, KaH., (2023), "Smart Heating, Ventilating, Air-conditioning and Refrigeration by Web-based Geographic Information System", Journal of Modern Industry and Manufacturing, 2, 6. https://doi.org/10.53964/jmim.2023006.
17. Sattar, A.M., and Chaudhry, M.H., (2008), "Leak Detection in Pipelines by Frequency Response Method", Journal of Hydraulic Research, 46(sup1), 138–151.
18. Duan, H.F., and Lee, P.J., (2016), "Transient-based Frequency Domain Method for Dead-end Side Branch Detection in Reservoir Pipeline-valve Systems", Journal of Hydraulic Engineering, 142(2), 04015042.
19. Hariri Asli H., and Nazari, S., (2021), "Water Age and Leakage in Reservoirs: Some Computational Aspects and Practical Hints", Larhyss Journal, 48, 151–167. http://larhyss.net/ojs/index.php/larhyss/article/view/808/807.
20. Haselbach, L., Adesina, M., and Muppavarapu, N. et al., (2023), "Spatially Estimating Flooding Depths from Damage Reports", Natural Hazards, 117, 1633–1645. https://doi.org/10.1007/s11069-023-05921-2.
21. Weidner, J., Collins, J., Benitez, M., Adesina, M., and Lozoya, C., (2019), Development of a Robust Framework for Assessing Bridge Performance Using a Multiple Model Approach, University of Texas at El Paso. Department of Civil Engineering, Report Number: CAIT-UTC-NC39. https://rosap.ntl.bts.gov/view/dot/48948.
22. Hariri Asli K., Hariri Asli H., and Motlaghzadeh, K., (2013), "Numerical Techniques in Water Transmission", Frontiers of Engineering Mechanics Research (FEMR), August, 2(3), 56–62, ISSN: 2306–6016 (Online), ISSN: 2306–6024 (Print), published by the world aca-demic publishing co., limited, Hong Kong, Corpus ID: 108917427. http://www.academicpub.org/femr/; https://api.semanticscholar.org/CorpusID:108917427.
23. Hariri Asli K., and Hariri Asli K., (2023). Minimum Night Flow (MNF) and Corrosion Control in Compliance with Internet of Things (IoT) for Water Systems. Water Practice and Technology. https://doi.org/10.2166/wpt.2023.012
24. Asli, KiH., and Asli, KaH., (2023), "Smart Water System and Internet of Things", Journal of Modern Industry and Manufacturing, 2, 5. https://doi.org/10.53964/jmim.2023005.

Implementation Water Loss by Smart Control Through the Internet of Things (IoT)

4

Abstract

Consumption management activities may change the time and size of new distribution network facilities such as sources and reservoirs, transmission and treatment facilities. Consumption management can save water and financial resources. The implementation of the water consumption management program leads to a reduction in costs. In this regard, it is necessary to use methods that reduce the need to create new facilities in the production sector. In some cases, the main costs related to the facilities cannot be avoided. Consumption management can save the repair and maintenance costs. In this work, the relationships between water consumption parameters and water production were investigated to evaluate the benefit and cost for consumption management programs. The amount of savings due to information and training program was 6 (%), Physical loss control was 29 (%), water reuse was 57 (%), and consumption reducers was 8(%). The results also showed the 30 (%) reduction of present Non-Revenue Water (NRW) by smart control through the Internet of Things (IoT).

Keywords

Non-Revenue Water (NRW) • Hydraulic model • Benefit and cost • District Metering Area • Water losses

4.1   Introduction

Consumption management is considered as a permanent strategy in determining the policy of development and operation of water and sewage infrastructure. When water distribution networks require the construction of new facilities, the benefits of consumption management increase significantly. By planning consumption management programs, it is possible not only to reduce the need for new water supply facilities that postpone the need for sewage facilities, or even eliminate them in some cases. Program designers should periodically review the goals of the program before and after the end of the consumption management program, because the goals and actions taken to achieve them have mutual relationships. When certain goals of the program are met, new goals may arise. Consumption management program is useful for most distribution networks. This benefit is not limited to networks that have upcoming investment projects. Even networks that estimate the capacity of their resources beyond their needs and facilities, realize that managing their consumption will help in the optimal use of resources and long-term savings in operating costs. As mentioned, eliminating, postponing or reducing the amount of investments related to new establishments is one of the main goals of the consumption management program.

So far, many efforts have been made to manage water consumption in different countries of the world. In order to face the challenges caused by water shortage, many cities in the world had to pay attention to comprehensive consumption management programs in both the supply and demand sectors [1–3].

In a research, 10,000 low-consumption toilets were distributed among about 10,000 consumers. A statistical comparison before and after the implementation of this plan for residential users showed that the average per capita consumption of each person in one day has decreased by 44 L. The total savings in scale. The whole network is equivalent to 37 L' reduction in per capita consumption per day. According to the number of consumers that participated in the project, 987 cubic meters of water were saved per day. More than 78% of the consumers who participated in the project were satisfied with the implementation of the project and the quality of the installed products. Also, the consumption management program brought great benefits to the society in environmental, economic and social fields. The number of jobs created by the consumption management plan increased. Also, in addition to saving in the consumption of available energies, the cost of the consumers' bills was reduced.

In another study, for a city that, in addition to facing a drought crisis, also faced the trend of increasing population, the water consumption management program including all the strategic measures required for a comprehensive plan was implemented. The main axes of the comprehensive plan were implemented as follows [3–6]:

- Comprehensive education program through advertising, publishing publications, setting up public workshops, training in schools, and setting up educational sites.
- Encouraging the use of equipment that improves consumption (consumption reduction).
- Legal support by setting standards for water consuming devices, drafting laws related to determining water prices.
- Research and help in the development of consumption improving equipment.

The main measures planned in the water consumption management program were as follows [7, 8]:

- Home consumption optimization program.
- Training program and improving the quality of home use of the yard and garden.
- Use of washing machines with optimal consumption and recovery equipment and use of rainwater.
- Optimization programs for commercial, industrial and government use.
- Real loss management programs including leak detection, quick and efficient repairs and pressure management.
- Water recycling and reuse programs in small home systems and large industrial systems.
- Legal amendments to benefit from water management methods.
- Amending the building regulations, amending the standards of building faucets and water consuming equipment.
- Developing regulations on the use of water in garden.

4.2   Materials and Methods

In this work, about 18% of household water was used in the yard and garden irrigation. The rain in the rainy season could satisfy a major part of this need. Household expenses related to shower, toilet and water faucets accounted for a major part of household expenses. The distribution of these equipment also caused a reduction in energy consumption due to savings in hot water consumption and a reduction in the production of pollutants and carbon dioxide during the research period.

Efficient equipment and connections were installed in the houses for optimal consumption. This optimization was accompanied by the installation of shower heads, flow regulating faucets, low-consumption siphons, and the elimination of drips, overflows, and internal leaks. This program was one of the largest programs for optimizing household consumption and distributing consumption-reducing equipment and covered about 15% of the citizens.

Implementation of the physical loss control program and leak detection on the distribution network saved money during the research period by IoT.

In addition, during the research period, the entire network was reviewed almost once. The main points of this program included the following:

- Active leakage management.
- Pressure management.
- Increasing the repair speed of the reported defects.
- Optimization of Minimum Night Flow ($M.N.F$) measurement systems.

4.2.1 Formulation

For District Metering Area (DMA) the Minimum Night Flow (*M.N.F*) was defined base on the 24 h' graph. The total flow entrance to the zone and Current Leakage Percentage (*C.L.P*) on water distribution network can be calculated by the following formulation (4.1–4.4) [9–11]:

(A) Minimum Night Flow (*M.N.F*) defined by 24 h' graph of *M.N.F*).
(B) Minimum Night Flow (*M.N.F*)

$$(\text{Convert item } A \text{ to } 1/c/h) = (M.N.F/C) \times 3600 \qquad (4.1)$$

(C) Legitimate Night Flow (*L.N.F*), (From Field Reading)
(D) Current Net Night Flow (*N.N.F*)

$$N.N.F = M.N.F(\text{item } B) - L.N.F(\text{item } C) \qquad (4.2)$$

(E) Total Leakage in zone, *L*

$$L = (N.N.F \times T \text{ FACTOR} \times C)/1000 \qquad (4.3)$$

(F) Total Flow to zone, *F* (Base on the 24 h' graph of *M.N.F*)
(G) Current Leakage Percentage (*C.L.P*)

$$C.L.P = L/F \times 100 \qquad (4.4)$$

4.2.2 Consumption Smart Management Program

Usually, the following basic steps are applied in the smart control through the Internet of Things (IoT) for consumption management program:

- Determining the objectives of the consumption management program.
- Methods of determining the existing situation.
- Methods of determining the demand function.
- Description of the planned facilities for the network in the program.
- Determining and introducing the necessary measures in the consumption management program.
- Benefit and cost analysis.
- Choosing the right actions.

- Integrating information sources and modifying production and demand functions.
- Determining implementation and evaluation strategies.

4.2.3 Cost-benefit and Consumption Management Plan

Performing a Cost-benefit analysis is optional. There are more complicated methods for predictions and analysis. In order to determine the demand and production function, analysis of cost and benefit and net profit resulting from various measures of the consumption management program and including more comprehensive measures are presented. Minimum measures depend on the size and capability of water production and distribution units and other conditions affecting these units such as climatic conditions, water availability and other effective factors. In other words, the basic actions of the consumption management program are not the same for small and large networks [12, 13].

The details of actions in each of the consumption management program levels are listed below:

4.2.3.1 Level 1 Measures

- Comprehensive measurement of flow in all points of the network.
- Water audit and loss control.
- Pricing and water tariff determination.
- Information and training.

4.2.3.2 Level 2 Measures

- Water consumption audit.
- Use of consumption reductions.
- Pressure management.
- Optimizing water consumption in green spaces.

4.2.3.3 Level 3 Measures

- Replacing and upgrading network components.
- Water reuse and recycling.
- Compilation of consumption regulations.
- Comprehensive resource management.

The most important and effective measures regarding consumption management are:

- The issue of flow measurement and meter repair.

- Leak detection and repair operations.
- Determining water rates according to the consumption management plan.
- Consumption reduction and promotion of consumption reduction devices.
- Public education and information.
- Irrigation efficiency of green spaces.

A complete consumption management program includes the following components [14, 15]:

4.2.4   Determining the Objectives of the Consumption Management Program

- The list of objectives of the consumption management program and their relationship with planning in the water supply sector.
- Explaining the way of public participation in the process of expanding the goals of the program.

4.2.5   Determining the Current Status of Distribution Networks.

- List of existing facilities, features of water supply and amount of consumption.
- An overview of the conditions affecting the distribution network and consumption management program.

4.2.6   Forecasting the Demand Function

- Prediction of future water needs for specific periods of time.
- Adjustment of needs based on known and measurable influencing factors.
- Discussion of uncertainties and sensitivity analysis (conditions and components).

4.2.7   Description of the Planned Facilities in the Program

- Improvement and correction programs for distribution networks in the horizon of the plan.
- Total, annual and unit cost estimation.
- The preliminary forecast of the overall capacity of water supply during the program is considered based on improvement and modification programs.

4.2.8   Introduction of Consumption Management Program Measures

- Examining the measures of the consumption management program that is being implemented or it has been decided to implement them.
- Examining the legal obstacles and other obstacles against the implementation of the recommended measures of the consumption management program.
- Complete identification of measures for more accurate analysis.

4.2.9   Cost Benefit Analysis

- Estimate the final cost of implementing the program and expected savings in water consumption.
- Carrying out a Cost-benefit evaluation for the recommended actions of the consumption management program.
- Comparison between the cost of implementing the program and the cost of savings.

4.2.10   Selection of Consumption Management Program Measures

- Selection criteria for choosing consumption management program measures.
- Introduction of selected actions.
- Provide appropriate explanations if the recommended actions are applicable.
- The strategy and timetable for the implementation of consumption management program measures.

4.2.11   Integrating Resources and Correcting Forecasts

- Modifying the predictions made for the amount of demand and production capacity in order to take into account the effects of the consumption management program.
- Explaining the effects of implementing the consumption management program on the amount of water consumed.

4.3   Presentation of Implementation and Evaluation Strategy

- Explaining the effects of implementing the consumption management program on the income of water supply and distribution units.

4.4   Results

The results of this work regarding the following items were included in Tables 4.1 and 4.2, Figs. 4.1 and 4.2.

Table 4.1 Amount of water saving due to consumption management programs

Type of action	Amount of savings (%)
Information and training	6
Physical loss control	29
Water reuse	57
Consumption reducers	8
Total	100

Table 4.2 Changes (%) due to consumption management Operations

Description of Operation	Changes(%)
Number of subscribers	0
Daily production of water (liters per day)	−48
Annual production of water (liters per year)	−48
Pumped water	−72
Working hours of pumping stations	−70
Water used for washing filters (liters)	−66
Non-Revenue Water (NRW) of present NRW	30

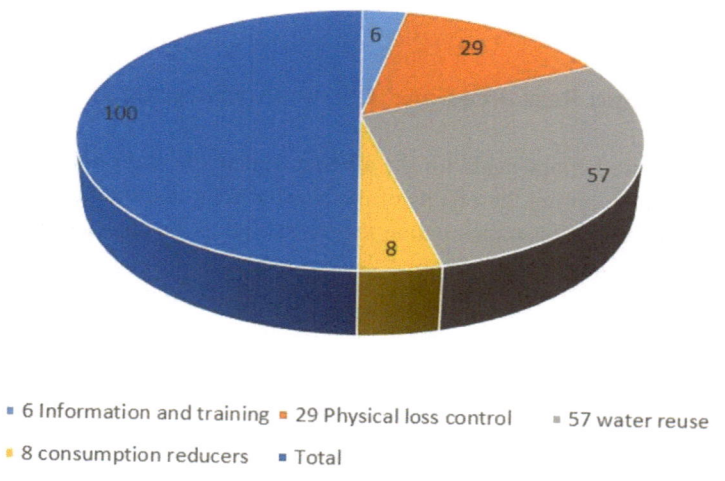

- 6 Information and training ▪ 29 Physical loss control ▪ 57 water reuse
- 8 consumption reducers ▪ Total

Fig. 4.1 3-D Pie graph for water saving due to consumption management programs

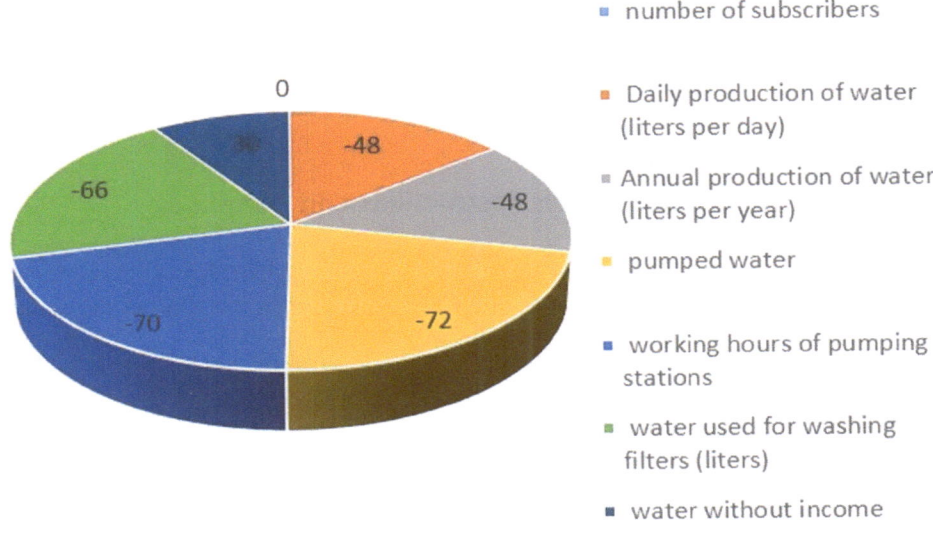

0

-48

-66   -48

-70   -72

- number of subscribers

- Daily production of water (liters per day)

- Annual production of water (liters per year)

- pumped water

- working hours of pumping stations

- water used for washing filters (liters)

- water without income

Fig. 4.2   3-D Pie graph for Changes (%) due to consumption management programs

- Information and training.
- Physical loss control.
- Gray water reuse.
- Consumption reducers.
- Number of consumers.
- Daily production of water (liters per day).
- Annual production of water (liters per year).
- Pumped water.
- Working hours of pumping stations.
- Water used for washing filters (liters).
- Non-revenue water (NRW).

In order to implement the programs as best as possible, the consumption management program emphasized on the training and precise guidance of the forces using the distribution networks. Some of the desired axes in order to achieve this goal were as follows [18–21].

4.4.1   Educational Courses and Workshops

Workshops and training courses are suitable tools for familiarizing managers, operators and executive forces with consumption management and the technical details of the planned actions. These courses and educational management program can be designed to familiarize water supply institutions with consumption laws. These courses focus on the process and steps of the planning process, such as demand forecasting and Cost-benefit analysis.

4.4.2 Preparation of Manuals, Instructions and Templates

Supplementary items such as instructions, templates, manuals, hypothetical examples and answers to the proposed questions can help the executive forces and users of distribution networks in facilitating the planning process.

4.4.3 Direct Personal Assistance

Providing direct expert assistance by phone and e-mail, especially for small networks that benefit from less technical power, can be very helpful. In this work, the preparation of labeling standards for water consuming equipment had an effect in terms of water consumption efficiency. This labeling included showers, pipes and fittings, siphon toilets, washing machines, laundry machines, overflow control floats and flow regulators.

4.5 Conclusion

Implementation of consumption smart management programs usually faces challenges. Traditional methods make it difficult to effectively use modern, intelligent and purposeful systems. In this work, the new experiences of the consumption management program emphasized the prominent role of saving in consumption. Also, the implementation of water consumption management showed that the use of modern and intelligent methods of consumption management improves service delivery to consumers. Finally, it has a positive effect on the growth of sales, income generation and profitability. Proposals for research projects in the field of water consumption management in the future are:

- Development of water consumption prediction models and demand modeling.
- Analysis of the consumption of smart buildings, checking the effect of the basic component of measuring the consumption of consumers.
- Investigating the quality of water in reservoirs collected from rain.
- Preparation of recycled water quality standards.
- Preparation of the necessary instructions for the make smart of water distribution in commercial and industrial units.
- Optimizing design standards for intelligent water distribution networks.

Abbreviations

$N.N.F$ Net Night Flow (From item D)
T Time Factor (From calculation of correction 'T' factor)
C No of connection

References

1. Haghighi, A., and Shamloo, H., (2011), "Transient Generation in Pipe Networks for Leak Detection", Proceedings of the Institution of Civil Engineers—Water Management, 164(6), 311–318.
2. Asli, KiH., and Asli, KaH., (2023), "Smart Heating, Ventilating, Air-conditioning and Refrigeration by Web-based Geographic Information System", Journal of Modern Industry and Manufacturing, 2, 6. https://doi.org/10.53964/jmim.2023006.
3. Soldevila, A., et al., (2017), "Leak Localization in Water Distribution Networks Using Bayesian Classifiers", Journal of Process Control, 55, 1–9.
4. Hariri Asli H., and Hozori, A., (2021), "Non-Revenue Water (NRW) and 3d Hierarchical Model for Landslide", Larhyss Journal, 48, 189–210. http://larhyss.net/ojs/index.php/larhyss/article/view/810/810.
5. Leu, S.-S., and Q.-N. Bui, (2016), "Leak Prediction Model for Water Distribution Networks Created Using a Bayesian Network Learning Approach", Water Resources Management, 30(8), 2719–2733.
6. Hariri Asli H., and Nazari, S., (2021), "Water Age and Leakage in Reservoirs: Some Computational Aspects and Practical Hints", Larhyss Journal, 48, 151–167. http://larhyss.net/ojs/index.php/larhyss/article/view/808/807.
7. Haghighi, A., Covas, D., and Ramos, H., (2012), "Direct Backward Transient Analysis for Leak Detection in Pressurized Pipelines: From Theory to Real Application", Journal of Water Supply: Research and Technology-Aqua, 61(3), 189–200.
8. Asli, H.H., (2023), Applications of Networked Sensors and the Internet of Things (IoT) for Water Treatment. Sustainable Water Treatment and Ecosystem Protection Strategies. Hard ISBN: 9781774915189. https://www.appleacademicpress.com/sustainable-water-treatment-and-ecosystem-protection-strategies-/9781774915189.
9. Asli, H.H., and Arabani, M., (2022), "Analysis of Strain and Failure of Asphalt Pavement. Computational Research Progress in Applied Science & Engineering", Transactions of Civil and Environmental Engineering, 8, 1–11. Article ID: 2250. https://doi.org/10.52547/crpase.8.1.2250.
10. Asli, H.H., (2023), Modeling of Corrosion for Water System by Networked Sensors and the Internet of Things (IoT) in Compliance with Geography Information System (GIS). Sustainable Water Treatment and Ecosystem Protection Strategies. Hard ISBN: 9781774915189. https://www.appleacademicpress.com/sustainable-water-treatment-and-ecosystem-protection-strategies-/9781774915189.
11. Asli, H.H., Arabani, M., and Golpour, Y., (2020), Reclaimed Asphalt Pavement (RAP) Based on a Geospatial Information System (GIS). Slovak Journal of Civil Engineering, 28(2), 36–42. https://doi.org/10.2478/sjce-2020-0013. Slovak University of Technology in Bratislava, Slovak.

12. Sarkamaryan, S., et al., (2020), "Surrogate-Assisted Inverse Transient Analysis (SAITA) for Leakage Detection in Pressurized Piping Systems", Iranian Journal of Science and Technology Transactions of Civil Engineering.
13. Haselbach, L., Adesina, M., and Muppavarapu, N., et al., (2023), "Spatially Estimating Flooding Depths from Damage Reports", Natural Hazards 117, 1633–1645. https://doi.org/10.1007/s11069-023-05921-2.
14. Weidner, J., Collins, J., Benitez, M., Adesina, M., and Lozoya, C., (2019), Development of a Robust Framework for Assessing Bridge Performance Using a Multiple Model Approach, University of Texas at El Paso. Department of Civil Engineering, Report Number: CAIT-UTC-NC39. https://rosap.ntl.bts.gov/view/dot/48948.
15. Hariri Asli K., Hariri Asli H., Motlaghzadeh, K., and Hariri Asli K., (2013), "Numerical Techniques in Water Transmission", Frontiers of Engineering Mechanics Research (FEMR), August, 2(3), 56–62, ISSN: 2306–6016 (Online), ISSN: 2306–6024 (Print), published by the world academic publishing co., limited, Hong Kong, Corpus ID: 108917427. http://www.academicpub.org/femr/; https://api.semanticscholar.org/CorpusID:108917427.
16. Hariri Asli K., and Hariri Asli K., (2022), Isolated Pressure Zones Based on GIS as a Solution for Water Network Problems. Water Practice and Technology. https://doi.org/10.2166/wpt.2022.119.
17. Sophocleous, S., Savić, D., and Kapelan, Z., (2019), "Leak Localization in a Real Water Distribution Network Based on Search-space Reduction", Journal of Water Resources Planning and Management, 145(7).
18. Hariri Asli K., Hariri Asli K., and Nazari, S., (2023), Computational Fluid Dynamics Analysis for Smart Control of Water Supply. Water Supply. https://doi.org/10.2166/ws.2023.306
19. Bohorquez, J., et al., (2020), "Leak Detection and Topology Identification in Pipelines Using Fluid Transients and Artificial Neural Networks", Journal of Water Resources Planning and Management, 146(6), 04020040.
20. Hariri Asli K., and Hariri Asli K., (2023), Minimum Night Flow (MNF) and Corrosion Control in Compliance with Internet of Things (IoT) for Water Systems. Water Practice and Technology. https://doi.org/10.2166/wpt.2023.012
21. Asli, KiH., and Asli, KaH., (2023), Smart Water System and Internet of Things. Journal of Modern Industry and Manufacturing, 2, 5. https://doi.org/10.53964/jmim.2023005.

Computational Modeling and Regression Analysis for Water Consumption Management

5

Abstract

Water consumption management and demand forecasting are prerequisites for evaluating the cost–benefit of implementing programs and understanding the challenges facing water systems. Forecasting the demand in the distribution networks of an isolated, sparsely populated area is considered one of the basic measures. During the water demand forecasting period, the water use efficiency program or public education plans are implemented to reduce water consumption. The results of this research showed that the forecasts will be more accurate when based on the classification of types of consumption. An implementation methodology is better to forecast the demand based on the supply population. This method will be suitable for networks that have small changes in the texture of the service-receiving population. Networks whose subscribers only include households in the residential sector and have fixed water consumption characteristics. In this method, per capita consumption is calculated. Per capita consumption is multiplied by the projected population. The predicted amount of water is compared with the capacity of the network. In this way, the surplus or shortage of predicted water is determined. The training process in the isolation area, showed that Consumption per capita is under the direct influence of General education. In the regression analysis, the fluid Consumption per capita (liters per day) as a dependent variable and the General education (h) as an independent variable had a P-value of 0.955.

Keywords

General education • Consumption per capita • Regression analysis • Isolated area • Water demand modeling

5.1   Introduction

Water demand prediction can be based on a simple basis such as population growth prediction and more complex models that operate based on various parameters. Forecasting
can be done on the entire distribution network and subscribers. Any type of adjustment
that takes place due to specific and known factors affecting consumption must be fully
explained. This method is used to estimate average daily consumption and maximum
daily consumption. An alternative for quick estimation of per capita water consumption
is to calculate water consumption per branch or each residential unit. For this purpose,
they usually use the average number of people living in each housing unit to convert into
calculations. To calculate water consumption, the per capita consumption of each residential unit is multiplied by the number of current and future branches. To predict future
residential units, one should be aware of the growth rate of construction [1–3].

Forecasting of water consumption of residential units is done accurately through per
capita consumption calculation methods. If the population and characteristics of the demo-
graphic structure are variable, separate forecasts should be made for major consumers. If a
major consumer tries to obtain a branch, or changes the type of branch, or waives the right
to branch, the effects on the network performance will be quite noticeable. The methods
of estimating per capita consumption and capita consumption of each housing unit have
limitations. Water consumption is a function of population or changing the number of
branches. The consumption pattern will not change with time. New technologies and the
installation of consumption-reducing equipment by subscribers affect changes in water
consumption. The introduction of new technologies, such as dishwashers with low water
consumption, attract the attention of customers.

For this purpose, it is necessary to prepare a general estimate of the following factors
that can affect the level and pattern of network consumption [4–6]:

- Per capita consumption parameter should be used for each branch.
- To determine the production capacity, the capacity of different sectors, such as the
  volume of available resources, the capacity of the distribution network, and the capacity
  of the treatment plants should be considered.

5.2   Materials and Methods

The implementation of water consumption management programs will reduce the amount
of water consumption or change in the predicted amounts of demand. Reducing water
consumption causes structural changes in consumption. In this research, an isolated area
was studied [7–12].

5.2.1 Population Estimation

There are many factors involved in estimating the population of an isolated area, the most important of which are:

- The growth trend of the city's population in the past years.
- Planned development plans for the future.
- The trend of population growth in nearby areas and nearby cities.
- Existing limitations against the future development of the city, such as the limitation of usable land and water resources.
- The possibility of assimilative migration or repulsive migration of the population.
- Seasonal population and population changes in different seasons of the year.

It should be mentioned that the population estimates are close to the reality if there are no changes in the structure and composition of the population. In order to predict the population, it is necessary to know the effect of the factors that change its number and structure and to diagnose it in a quantitative way for certain futures. Considering the many factors involved in population growth, it is very difficult to predict and estimate the working population. None of the formulas and methods provided for population prediction include all the effective factors in population growth. Therefore, it is not possible to rely only on a specific formula or method. Anyway, by conducting comprehensive studies and using demography experts, it is possible to increase the accuracy of the results related to population forecasts. There are different methods to predict the population, which can be divided into four categories and the following methods [13–15]:

- Graphical methods.
- Mathematical methods.
- Ratio and correlation methods.
- Component methods.
- Combined methods.

In general, there are two types of population forecasting:

- Short-term estimate for 1 to 10 years.
- Long-term forecast for 10 to 50 years or more.

The methods used for these two types of population forecasting are different. Graphical and mathematical methods can be used for short-term population estimation. Short-term population growth forecasting methods include two types of geometric growth and arithmetic growth. Each part has an independent mathematical relationship. In order to find the appropriate mathematical relationship, the past data of the studied area should be entered on a population-time coordinate system and their curves should be drawn [16, 17].

5.2.2   Arithmetic Growth Method

In this method, which is also known as the uniform rate method, it is assumed that the population increase dp in equal times dt is constant and independent of the population (5.1–5.2).

$$dP/dt = K_a \qquad (5.1)$$

$$K_a = (P_2 - P_1)/(T_2 - T_1) \qquad (5.2)$$

K_a is obtained by integrating formula (5.1). As a result, the population is estimated from the following relationship (5.3):

$$P_t = P_0 + K_a t \qquad (5.3)$$

where P_t is the predicted population in t year after. P_0 is the current population.

5.2.3   Geometric Growth Method

In this method, which is also known as the uniform percentage growth method or the logarithmic method, it is assumed that in equal periods, the percentage of constant growth or, in other words, the rate of increase of the population is proportional to the population (5.4–5.6) [18–27].

$$(dP/dt) = K_g P \qquad (5.4)$$

$$K_g = (LnP_2 - LnP_1)/(T_2 - T_1) \qquad (5.5)$$

$$LnP_t = (LnP_0 - K_g t) \qquad (5.6)$$

Fig. 5.1 Pie chart for changes in per capita water consumption versus time

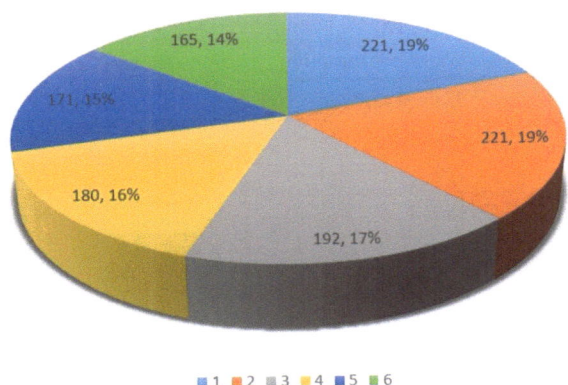

Table 5.1 Changes in per capita water consumption versus time

No.	Consumption per capita (liters per day)	Year
1	221	2015
2	221	2016
3	192	2017
4	180	2018
5	171	2019
6	165	2020

5.3   Results

In this work, changes in water consumption per capita with time have been investigated. The amount of water consumption per capita fluctuates significantly with time. To estimate the general trend of this characteristic, the six-year moving average values of the data have also been plotted (Fig. 5.1, Table 5.1).

Knowing the behavior of the dependent variable (water consumption per capita) with time, various curves such as linear, logarithmic, exponential, and exponential curves have been fitted to the data in regression analysis. The population of the isolated area was estimated. The use of public education, low-volume toilet siphons, and the use of green plants that need less water were investigated. The independent variables of the research include public education, low-volume toilet siphons, and green plants with low water consumption. Per capita water consumption as a dependent variable of the research was included in the regression analysis for water demand predictions (Tables 5.2, 5.3 and 5.4, Fig. 5.2).

Table 5.2 Changes of water consumption versus General education

No.	Water consumption per capita (liters per day)	Year	General education (h)
1	221	2015	120
2	221	2016	200
3	192	2017	220
4	180	2018	400
5	171	2019	400
6	165	2020	550

Table 5.3 Model Summary and Parameter Estimates Dependent Variable: Consumption per capita (liters per day); The independent variable is General education (h)

Equation	Model summary					Parameter estimates			
	R Square	F	df_1	df_2	Sig.	Constant	b_1	b_2	b_3
Linear	0.845	21.870	1	4	0.009	235.638	−0.140		
Logarithmic	0.864	25.445	1	4	0.007	418.112	−40.230		
Inverse	0.791	15.157	1	4	0.018	154.619	9000.549		
Quadratic	0.876	10.590	2	3	0.044	255.048	−0.286	0.000	
Cubic	0.879	4.825	3	2	0.176	240.782	−0.122	0.000	5.19E-007
Compound	0.863	25.245	1	4	0.007	239.430	0.999		
Power	0.872	27.365	1	4	0.006	615.678	−0.209		
S	0.790	15.063	1	4	0.018	5.058	46.396		
Growth	0.863	25.245	1	4	0.007	5.478	−0.001		
Exponential	0.863	25.245	1	4	0.007	239.430	−0.001		
Logistic	0.863	25.245	1	4	0.007	0.004	1.001		

Table 5.4 Statistics Test for Consumption per capita (Dependent Variable) and General education (Independent Variable)

Parameters	Consumption per capita (liters per day)	General education (h)
Chi-Square	0.667	0.667
df	4	4
Asymp. Sig.	0.955	0.955

Consumption per capita (liters per day)

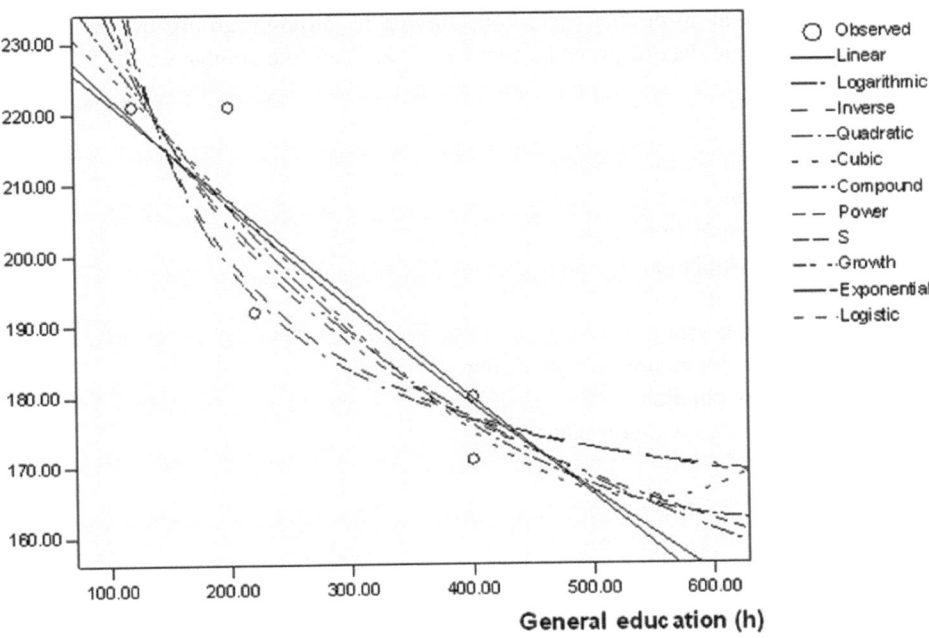

General education (h)

Fig. 5.2 Scatter diagram for Dependent Variable: Consumption per capita (liters per day); The independent variable is General education (h)

5.4 Conclusion

An implementation methodology is better to forecast the demand based on the supply population. This method will be suitable for networks that have small changes in the texture of the service-receiving population. Networks whose subscribers only include households in the residential sector and have fixed water consumption characteristics. The results of this work are as follows:

- Public education plans are implemented to reduce water consumption.
- Consumption per capita is under the direct influence of General education.
- Forecasts will be more accurate when based on the classification of types of consumption.
- An implementation methodology is better to forecast the demand based on the supply population.

Suggestions for Future Research

In this work, a method was used to predict water consumption per capita, which can be used in a similar way to predict water consumption by different groups of users. In this method, the available data of previous years are examined in a similar way.

Abbreviations

K_a	Arithmetic growth constant
P	Population
P_1	Census population in year T_1
P_2	Census population in year T_2
T_1	Current year
T_2	Destination year
P_t	Predicted population in t year after
P_0	Current population
K_g	Geometric growth constant

References

1. Barnett, M.J., (2018), A Multilevel Analysis of Social, Built, and Natural Drivers of Household Water Use in Northern Utah (Doctoral dissertation, The Ohio State University).
2. Gondo, R., and Kolawole, O.D., (2019), "Sustainable water resources management: issues and principles of water governance in the Okavango Delta, Botswana", International Journal of Rural Management, 15(2), 198–217.
3. Hariri Asli, H, Hozori, A., (2021), Non-Revenue Water (NRW) and 3d Hierarchical Model for Landslide. Larhyss Journal, 48, 189–210. http://larhyss.net/ojs/index.php/larhyss/article/view/810/810.
4. Zhang, X., Srinivasan, R., and Van Liew, M., (2009), "Approximating SWAT Model Using Artificial Neural Network and Support Vector Machine", JAWRA Journal of the American Water Resources Association, 45(2), 460–474.
5. Hariri Asli, H., and Nazari, S., (2021), "Water Age and Leakage in Reservoirs: Some Computational Aspects and Practical Hints", Larhyss Journal, 48, 151–167. http://larhyss.net/ojs/index.php/larhyss/article/view/808/807.
6. Tao, T., et al., (2014), "Burst Detection Using an Artificial Immune Network in Water-Distribution Systems", Journal of Water Resources Planning and Management, 140(10), 04014027.
7. Asli, H.H., (2023), Applications of Networked Sensors and the Internet of Things (IoT) for Water Treatment. Sustainable Water Treatment and Ecosystem Protection Strategies. Hard ISBN: 9781774915189. https://www.appleacademicpress.com/sustainable-water-treatment-and-ecosystem-protection-strategies-/9781774915189.
8. Asli, H.H., and Arabani, M., (2022), "Analysis of Strain and Failure of Asphalt Pavement. Computational Research Progress in Applied Science & Engineering", Transactions of Civil and Environmental Engineering, 8, 1–11. Article ID: 2250. https://doi.org/10.52547/crpase.8.1.2250.

9. Mounce, S.R., Mounce, R.B., and Boxall, J.B., (2011), "Novelty Detection for Time Series Data Analysis in Water Distribution Systems Using Support Vector Machines", Journal of Hydroinformatics, 13(4), 672–686.

10. Chaudhry, M.H., (2016), Applied Hydraulic Transients. Springer, New York.

11. Asli, H.H., (2023), Modeling of Corrosion for Water System by Networked Sensors and the Internet of Things (IoT) in Compliance with Geography Information System (GIS). Sustainable Water Treatment and Ecosystem Protection Strategies. Hard ISBN: 9781774915189. https://www.appleacademicpress.com/sustainable-water-treatment-and-ecosystem-protection-strategies-/9781774915189.

12. Asli, H.H., Arabani, M., and Golpour, Y., (2020), Reclaimed Asphalt Pavement (RAP) Based on a Geospatial Information System (GIS). Slovak Journal of Civil Engineering, 28(2), 36–42. https://doi.org/10.2478/sjce-2020-0013. Slovak University of Technology in Bratislava, Slovak.

13. O'Leary, H., (2019), Conspicuous Reserves: Ideologies of Water Consumption and the Performance of Class. Economic Anthropology.

14. Hariri Asli, K., Hariri Asli, H., Motlaghzadeh, K., and Hariri Asli, K., (2013), "Numerical Techniques in Water Transmission", Frontiers of Engineering Mechanics Research (FEMR), August, 2(3), 56–62, ISSN: 2306–6016 (Online), ISSN: 2306–6024 (Print), published by the world academic publishing co., limited, Hong Kong, Corpus ID: 108917427. http://www.academicpub.org/femr/; https://api.semanticscholar.org/CorpusID:108917427.

15. Hariri Asli, K., and Hariri Asli, K., (2022), Isolated Pressure Zones Based on GIS as a Solution for Water Network Problems. Water Practice and Technology. https://doi.org/10.2166/wpt.2022.119.

16. Mashford, J., et al. (2009), An Approach to Leak Detection in Pipe Networks Using Analysis of Monitored Pressure Values by Support Vector Machine. 2009 Third International Conference on Network and System Security.

17. Zhang, Q., et al. (2016), "Leakage Zone Identification in Large-scale Water Distribution Systems Using Multiclass Support Vector Machines", Journal of Water Resources Planning and Management, 142(11), 04016042.

18. Haselbach, L., Adesina, M., and Muppavarapu, N., et al., (2023), "Spatially Estimating Flooding Depths from Damage Reports", Natural Hazards, 117, 1633–1645. https://doi.org/10.1007/s11069-023-05921-2.

19. Weidne, J., Collins, J., Benitez, M., Adesina, M., and Lozoya, C., (2019), Development of a Robust Framework for Assessing Bridge Performance Using a Multiple Model Approach, University of Texas at El Paso. Department of Civil Engineering, Report Number: CAIT-UTC-NC39. https://rosap.ntl.bts.gov/view/dot/48948.

20. Hariri Asli, K., Hariri Asli, K., and Nazari, S., (2023), Computational Fluid Dynamics Analysis for Smart Control of Water Supply. Water Supply. https://doi.org/10.2166/ws.2023.306.

21. Prakash, A., Singh, S., and Brouwer, L., (2015), "Water Transfer from Peri-urban to Urban Areas: Conflict Over Water for Hyderabad City in South India", Environment and Urbanization ASIA, 6(1), 41–58.

22. Hariri Asli, K., and Hariri Asli, K., (2023), Minimum Night Flow (MNF) and Corrosion Control in Compliance with Internet of Things (IoT) for Water Systems. Water Practice and Technology. https://doi.org/10.2166/wpt.2023.012.

23. Asli, KiH., and Asli, KaH., (2023), "Smart Water System and Internet of Things", Journal of Modern Industry and Manufacturing, 2, 5. https://doi.org/10.53964/jmim.2023005.

24. Walt, J.C., Heyns, P.S., and Wilke, D.N., (2019), "Pipe Network Leak Detection: Comparison Between Statistical and Machine Learning Techniques", Urban Water Journal, 5(10), 953–960.

25. Bergua Amores, J.A., (2008), "Ideology, Magic and Spectres: Towards a Cultural Analysis of Water Use", Current Sociology, 56(5), 779–797.

26. Asli, KiH., Asli, KaH., (2023), "Smart Heating, Ventilating, Air-conditioning and Refrigeration by Web-based Geographic Information System", Journal of Modern Industry and Manufacturing, 2, 6. https://doi.org/10.53964/jmim.2023006.
27. Phipps, M., and Brace-Govan, J., (2011), "From Right to Responsibility: Sustainable Change in Water Consumption", Journal of Public Policy & Marketing, 30(2), 203–219.

Advanced Technologies and Forecasting Models for Water Demand

6

Abstract

Prediction of water consumption or demand is an important part of the planning process in terms of consumption management plans. Predictions range from simple estimates based on population growth to Predictions based on complex models that consider different variables in water consumption, and the change the predictions can be made for the whole network. the predictions for different groups of consumers (residential, commercial, public) are more accurate. In general, water demand forecasting methods are divided into different forms. Sometimes these methods are combined with new methods. Hence the clear division It is difficult to determine a clear border between them. To design different urban water systems such as wells, pumps, distribution networks, treatment plants, and reservoirs, forecasting water demand in different periods is needed. Demand forecasting is usually classified into three groups: short-term, medium-term, and long-term forecasting. Usually, forecasts of one to two years, short-term forecasts, between one to ten years, and mid-term and longer periods are classified as long-term forecasts. The water demand forecast investigated consumption (m^3/year) in the isolation area. The results of research showed that Consumption is under the direct influence of NRW (%). In the regression analysis, the water demand forecast as a dependent variable and the NRW as an independent variable had a P-value of 0.955.

Keywords

Water consumption forecast • Water demand forecast • Water consumption model • Isolated area • Consumption management

6.1   Introduction

Short-term forecasts are usually made to allocate budget and financial planning shortly time. Financial planning includes income from water sales and network costs. through predictions, the medium term planning and developing can be used for the distribution network and purification systems. In these forecasts, changes in water consumption due to changes in the project conditions such as water and air changes, Industrial and commercial are also considered, based on these studies, the water rate can be determined. Long-term forecasts are also used for financial planning for development in the distant future. Also, in these forecasts, more attention is paid to the changes in the water supply system, including the existence of water resources, fluctuations in resources, quality, and time changes in water demand. Based on the results of these forecasts, the development policies and determination of water allocation priorities will be determined to realize the sustainable and long-term capacity of water resources. As the forecasting time increases, the error in estimating water needs increases. Predictions can be made for five-year, ten-year, and twenty-year intervals. Can be from distances He used it another time, but the longer the planning horizon, the greater the uncertainty of the predictions became. Forecasts should be reviewed and updated at specific times. Designers should also estimate the average and maximum daily consumption during the planning horizon because depending on the type of facilities, the production and distribution sectors are designed to meet the peak water demand or the average demand, and the consumption management plan measures can be used [1–3].

6.2   Materials and Methods

Among the basic concepts for forecasting are curve fitting and prediction of results for the future period. According to this concept, it is assumed that the trend of variable parameter changes, such as the prediction of water consumption in the future, is also by the past trend. To determine water consumption, the curve fitting method can be used directly to determine the amount of water consumed per capita or for each user. In another method, the relationship between different characteristics (which are called independent variables) and the amount of water consumption (which is the dependent variable) are determined first, and a relationship between the dependent variable and the independent variables is extracted. Now, to predict the amount of water consumption, it is necessary to determine the independent variables for the period of the forecast plan and then based on the extracted relationship, the amount of water consumption is determined. This method can be used to predict consumption per capita, household, or consumption of each branch. The data used in this method can be case or time series. In the first case, it is possible to predict annual consumption, and in the second case, monthly consumption for different months [4, 5]. Among other forecasting methods is the final consumption method. In this method, first, how water is consumed in different parts is investigated. For example, if we know that one of the components of water consumption is a dishwasher and we know that a dishwasher uses an average of one thousand cubic meters of water per year, by predicting the number of dishwashers, can be total He estimated the annual consumption. In this way, by identifying all the consumption components and predicting their amount, the amount of water consumption can be estimated. Also, each of the consumption management measures should be evaluated and analyzed for sensitivity before changes. The important and basic ones that are created in the level and pattern of consumption should be identified. Predictions and plans should explain and calculate any significant changes in consumption [6, 7].

6.2.1 Weather Conditions

Climatic classification is a description of the weather condition of a region and it has mostly a conversational aspect. Many of the climatic classification methods are experimental or proposed. The simplest and most elementary classification of climates is based on air temperature. In this system, there are three distinct types of climates in the world [8, 9]:

- Tropical weather
- Mild climate
- Polar weather

Tropical climate does not have a cold season. The polar climate does not have a hot season.
There are different systems and classifications for climate, which include:

6.2.2 Coupon Method

The coupon method is based on the relationship between rainfall and temperature. Based on Kopan's comments, there are three types of climates, which are desert climates, steppe climates, and humid climates, which are given below the simplified Kopan plan. Three types of climate tes based on Kopan's comments.

6.2.3 Dumarten's Method

Dumarten's method is based on temperature and humidity and the following relationship (6.1):

$$I_{dc} = P_{aar}/(T_{aa} + 10) \tag{6.1}$$

Based on this, 6 types of climates can be determined, which are in the range of $10 \leq I < 35$.

The name of the climate of the dry area of Dumaraton

- Dry ($I < 10$).
- semi dry ($10 < I < 19.9$).
- Mediterranean ($20 < I < 23.9$).
- semi moist ($20 < I < 27.9$).
- wet ($28 < I < 34.9$).
- Very humid ($I > 35$).

6.2.4 Ivanov's Method

Ivanoff's method is based on the amount of rainfall and evaporation. In this method, the humidity coefficient (I) is obtained from the following relations (6.2, 6.3):

$$I_{Ihc} = P_{aar}/\sum E \tag{6.2}$$

$$E = 0.0018(2.5 + T2)(100 - r) \tag{6.3}$$

Based on this, the whole climate can be divided into six categories.
The type of climate, the range of Ivanoff's humidity coefficient

- Very humid forest areas ($I < 1.5$).
- Wet forest areas ($1 < I < 1.49$).
- Forest steppe areas ($16 < I < 0.99$).
- Steppe ($0.13 < I < 0.59$).
- Desert ($0.13 < I < 0.29$).
- Desert ($0 < I < 0.12$).

6.2.5 Barat Method

Barat method is divided based on the severity of dryness. The climate coefficient is obtained using the following relationship (6.4):

$$I_{bcc} = (P_{aar}(1 - C)/(N - 365)) - E/365 \tag{6.4}$$

Therefore, the whole climate is divided into four regions. The name of the climate is Zarib Barat.

- Desert (I < -20).
- semi dry ($-20 < I < 0$).
- semi moist ($0 < I < 7$).
- Forest humid ($7 < I$).

6.2.6 Trent White Method

Trent White's method is based on evaporation and perspiration. In the Trent White method, the effective rain profile is obtained from the following relationship (6.5):

$$P_{aar}E_{ae}I_{dc} = 115[P_{amr}/(t - 10)]1.11 \qquad (6.5)$$

Therefore, the whole climate is divided into five types of regions.
PEI climate name

- Dry (PEI < 16).
- Semi dry (16 < PEI < 31).
- Semi moist (32 < PEI < 63).
- Wet (64 < PEI < 129).
- Very wet (128 < P.E.I).

6.2.7 Selaninov's Method

Selaninov's method is based on the ratio of moisture to heat. The so-called hydrothermal Selaniov coefficient is obtained from the following Eq. (6.6):

$$C = \sum P_{tap} / \left(0.10 \sum H \right) \qquad (6.6)$$

6.2.8 Water Loss Calculation (6.7)

$$\text{Water Loss} = (\text{Quantity of water put in to supply}$$
$$- (\text{Non-domestic usage} + \text{Domestic consumption})) \qquad (6.7)$$

6.2.9 Non-Revenue Water (NRW) (6.8)

$$U = S - \left(M + \left(A * P_{ps}\right)\right) \qquad (6.8)$$

6.2.10 Non-Revenue Water (NRW) in the Distribution Network

Non-Revenue Water (NRW) in the water distribution network includes the following situations [10–12]:

- Water losses due to fractures and incidents (visible leakage).
- Water losses due to small leaks (invisible leaks).
- Water losses from faucets and fittings.
- Removal of fire extinguishers.
- Washing the network through the available drain valves.

Small leaks caused by faucets and fittings are considered to be part of small or invisible leaks. In this way, the minimum amount of water not taken into account due to faucets and fittings can be calculated. Since leaks caused by broken pipes and connections in the water distribution network are identified and fixed after receiving a leak report from residents, leaks caused by network failures and incidents are considered visible types of leaks. Regarding the leakage caused by the removal of fire hydrants, according to the investigations, this consumption is supplied through the wells that have been allocated for this purpose, and it has not been included in the calculations of the leakage from the network [13, 14].

6.2.11 Water Balance Method of Computing Leakage

Computation of Leakage Water Balance Formula is as the following (6.9):

$$U = S - \left(M + A * P_{psu}\right) \qquad (6.9)$$

The background night flow losses when no bursts exist in a District Metering Area (DMA) can be calculated for any DMA; given from the Eq. (6.10):

$$\text{NFLB}\left(\frac{l}{\text{hr}}\right) = (C1 * L + (C2 + C3) * N) * \text{PCF} \qquad (6.10)$$

Using the following values of C1, C2 and C3 from Table 6.1, and the Pressure Correction Factor (PCF) (Tables 6.1 and 6.2).

Table 6.1 Night Flow Losses Background (NFLB)

Background Losses Component	Units	Low	Average	High
C1	L/KM/HR	20	40	60
C2	L/PROP/HR	1.5	3	4.5
C3	L/PROP/HR	0.5	1	1.5

Table 6.2 The values of Pressure correction factors (PCF)

AZPN (meters)	20	30	40	50	60	70	80	90	100
PCF	0.329	0.529	0.753	1	1.271	1.565	1.884	2.226	2.592

6.2.12 Demand Function

We consider the demand function as follows (6.11):

$$Q = f(P, Z) \tag{6.11}$$

P is the value and Z is the vector of other independent factors.

In some studies, the size of the garden of the house, the number of bathrooms, and sanitary services are desired [6, 15].

- Frequency of bill payment and design of tariff system.
- There is a positive relationship between the frequency of bill payments and the amount of consumption, and the rising tariff system can also cause a logical reduction in consumption.
- Internal expenses, external expenses, and seasonal expenses are among the indicators and variables.
- The price of other goods is important to determine the nature of the goods from the perspective of substitution or complementarity and the effect of the increase in the price of other goods on water consumption.
- Income: The income criterion is usually obtained from the total income of the isolated area divided by the number of population or households.
- An increase in income causes a change in a person's consumption preferences. A variable such as the number of people with
- Household income, the number of cars, is considered as a substitute for income.
- Weather variables: the effect of variables such as average monthly temperature and average rainfall.

Table 6.3 Changes in water
consumption (m^3 per year)
versus NRW (%)

No.	Consumption (m^3 per year)	NRW (%)
1	79.56	32
2	79.56	32
3	69.12	29
4	64.80	28
5	61.56	25
6	59.40	22

The formulation of demand functions is usually obtained from the maximization of the utility function concerning the budget line. Mostly, the demand function is considered linear. The problem that exists in such linear demand functions is that the price elasticity is not the same at all points of the demand function. Usually, to obtain elasticity in these functions, the average value and the average price are used, and if uniform prices are used, in contrast to high or very low prices, the reaction of consumers is not observed. Of course, in double logarithmic functions, the strains can be calculated from the function itself, and the coefficients of the independent variables of the strain are independent variables relative to the value. In all points, this stretch is the same and it is considered a weak point. It is not different for high and low prices. Below, an important demand equation that has many applications is reviewed [16–26].

6.2.13 Nags and Thomas Equation

The model of Nags and Thomas is as follows (6.12):

$$C_{it} = x_{it}\beta + Z_i^c \gamma + \alpha_i + \eta_{it} \tag{6.12}$$

In the regression analysis of the demand function of Nags and Thomas, the parameter αi under the title of NRW has a direct effect on the demand function.

6.3   Results

In the regression analysis of the demand function of Nags and Thomas, the amount of NRW is the independent variable and the amount of water demand is the dependent variable. The present study investigated the relationship between the amount of water demand in an isolated area as a dependent variable and the amount of NRW as an independent variable (Tables 6.3, 6.4 and 6.5, Figs. 6.1, 6.2 and 6.3).

Table 6.4 Model summary and parameter estimates Dependent Variable: Consumption (m^3/year) and the independent variable is NRW (%)

Equation	Model summary					Parameter estimates			
	R Square	F	df$_1$	df$_2$	Sig.	Constant	b$_1$	b$_2$	b$_3$
Linear	0.894	33.703	1	4	0.004	9.942	2.109		
Logarithmic	0.858	24.208	1	4	0.008	−116.068	55.685		
Inverse	0.818	18.013	1	4	0.013	121.539	−1445.028		
Quadratic	0.994	244.059	2	3	0.000	180.714	−10.648	0.234	
Cubic	0.995	285.463	2	3	0.000	86.463	0.000	−0.163	0.005
Compound	0.913	41.889	1	4	0.003	29.145	1.031		
Power	0.880	29.311	1	4	0.006	4.675	0.808		
S	0.842	21.389	1	4	0.010	4.991	−21.008		
Growth	0.913	41.889	1	4	0.003	3.372	0.031		
Exponential	0.913	41.889	1	4	0.003	29.145	0.031		
Logistic	0.913	41.889	1	4	0.003	0.034	0.970		

Table 6.5 Statistics test for Dependent Variable: Consumption (m^3/year) and the independent variable is NRW (%)

Parameters	Consumption (m^3/year)	NRW (%)
Chi-Square	0.667	0.667
df	4	4
Asymp. Sig.	0.955	0.955

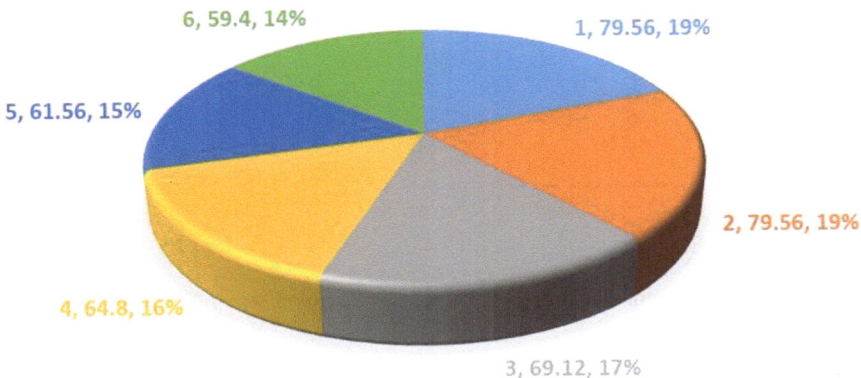

CONSUMPTION (M3 PER YEAR)

6, 59.4, 14%
5, 61.56, 15%
4, 64.8, 16%
3, 69.12, 17%
2, 79.56, 19%
1, 79.56, 19%

Fig. 6.1 Changes in water Consumption (m^3 per year)

Fig. 6.2 Changes in NRW (%)

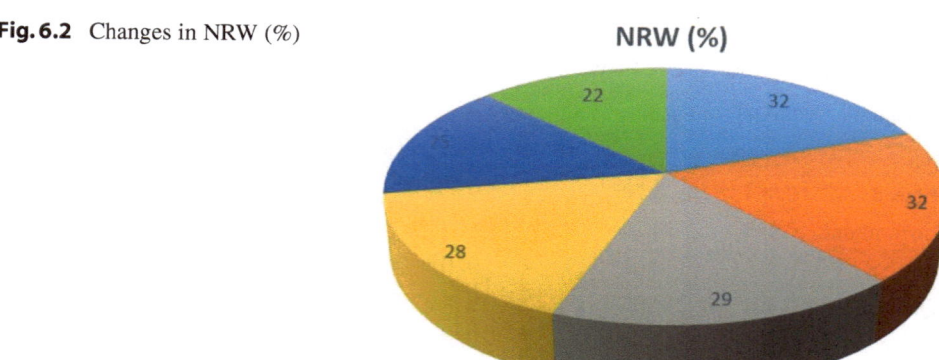

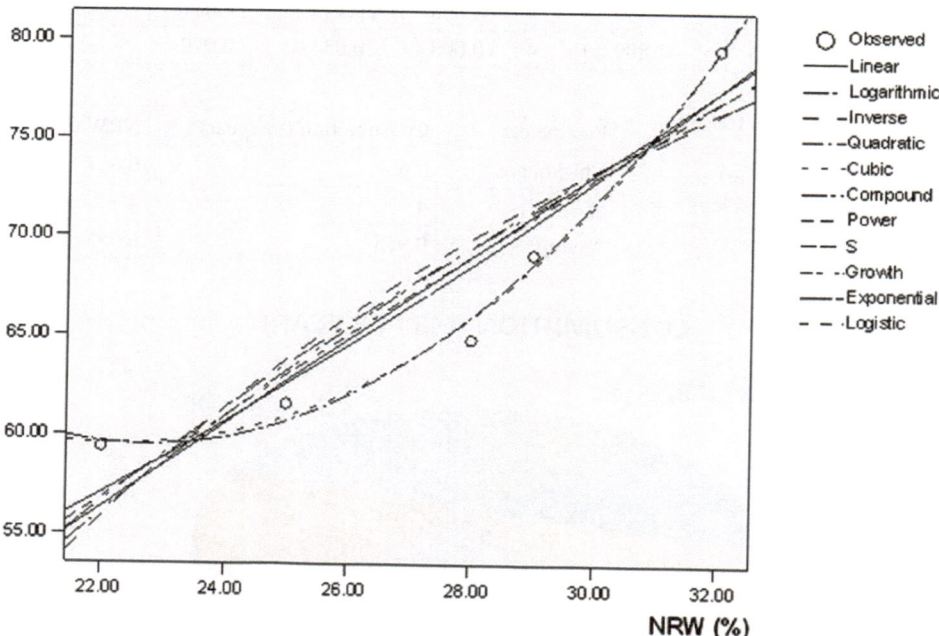

Fig. 6.3 Scatter diagram for Dependent Variable: Consumption (m³/year) and The independent variable is NRW (%)

6.4 Conclusion

To develop the model, it is necessary for the designer to have a conceptual view of the variables that can be used and their relationship with each other. The results of this work are as follows:

- Demand forecasting is usually classified into three groups: short-term, medium-term, and long-term forecasting. Usually, forecasts of one to two years, short-term forecasts, between one to ten years, and mid-term and longer periods are classified as long-term forecasts.
- In the regression analysis, the water demand forecast as a dependent variable and the NRW as an independent variable had a P-value of 0.955.
- In the regression analysis of the demand function of Nags and Thomas, the parameter αi as an NRW indicator has a direct effect on the demand function.

Suggestions for Future Research
In this work, a method was used to predict water demand per capita, which can be applied in a similar way to predict water demand by different groups of users. In this method, the water demand per capita as a dependent variable and the NRW as an independent variable was studied. for future research, the study can be carried out on water demand for all consumers.

Abbreviations

K_a	Arithmetic growth constant
P_1	Census population in year T_1
P_2	Census population in year T_2
P_t	Predicted population in year t after P_0
P_0	Current population
P	Annual rainfall amount (cm)
T_{aa}	Average annual temperature (°C)
I_{dc}	Dry coefficient
P_{aar}	Average annual rainfall (mm)
I_{lhc}	Ivanoff humidity coefficient
T_{am}	Average monthly temperature (°C)
R	Average monthly relative humidity (%)
E_{me}	Monthly evaporation (cm)
$\sum E$	Sum of evaporation in the months of the year (cm)
I_{bcc}	Barat climatic coefficient
C	Surface runoff coefficient
N	Number of rainy days per year

E_{ae}	Annual evaporation (mm)
P_{amr}	Average monthly rainfall (m)
T	Average monthly temperature (°C)
$\sum P_{tap}$	The total amount of precipitation (cm) in a period of time in which the average temperature is higher than 10 (°C)
$\sum H$	Cumulative value of temperature in the same time period (°C)
U	Non-Revenue Water (NRW) for quantities of water including leakage
S	Sum of all water inputs into a system
M	Sum of all water accounted for by measure (metered supplies, domestic and non-domestic)
A	Average domestic usage per capita of population
P_{ps}	Population supplied (non-metered)
PCF	Pressure Correction Factor
U	Non-Revenue Water (NRW)
S_{tvs}	Total volume supplied
M	Metered use
A	Per capita use
P_{psu}	Population supplied unmetered
NFLB	Night Flow Losses Background
L	Length of mains (km)
N_{nop}	Number of properties
AZNP	Average Zonal Night Pressure (m)
C_{it}	Amount of water demand in the area
x_{it}	Price vector
$Z_i{}^c$	Vector of socio-economic variables
α_i	NRW level
η_{it}	Remaining values
Z	Vector of other independent factors
Q	Demand (m³/year)
P	Price (Dollar)

References

1. Lambrinou, C.P., van Stralen, M.M., Androutsos, O., Moreno, L.A., Iotova, V., Socha, P., and Manios, Y., (2018), Mediators of the Effectiveness of an Intervention Promoting Water Consumption in Preschool Children: The ToyBox Study. Journal of School Health, 88(12), 877–885.
2. Asli, H.H., and Hozori, A., (2021), Non-Revenue Water (NRW) and 3d Hierarchical Model for Landslide. Larhyss Journal, 48, 189–210. http://larhyss.net/ojs/index.php/larhyss/article/view/810/810.

3. Choi, I.C., (2017), "Water Policy Reforms in South Koea: A Historical Review and Ongoing Challenges for Sustainable Water Governance and Management", Water 9(9), 717.

4. Asli, H.H., and Nazari, S., (2021), "Water Age and Leakage in Reservoirs: Some Computational Aspects and Practical Hints", Larhyss Journal, 48, 151–167. http://larhyss.net/ojs/index.php/larhyss/article/view/808/807.

5. Ishiwatari, M., and Song, Y., (2017), Promoting Green Growth Through Water Resources Management: The Case of Republic of Korea. Water Global Practice Group, World Bank Group.

6. Haselbach, L., Adesina, M., and Muppavarapu, N. et al., (2023), "Spatially Estimating Flooding Depths from Damage Reports", Natural Hazards, 117, 1633–1645. https://doi.org/10.1007/s11069-023-05921-2.

7. Weidner, J., Collins, J., Benitez, M., Adesina, M., and Lozoya, C., (2019), Development of a Robust Framework for Assessing Bridge Performance Using a Multiple Model Approach, University of Texas at El Paso. Department of Civil Engineering, report number: CAIT-UTC-NC39. https://rosap.ntl.bts.gov/view/dot/48948.

8. Asli, H.H., (2023), Applications of Networked Sensors and the Internet of Things (IoT) for Water Treatment. Sustainable Water Treatment. and Ecosystem Protection Strategies. Hard ISBN: 9781774915189. https://www.appleacademicpress.com/sustainable-water-treatment-and-ecosystem-protection-strategies-/9781774915189.

9. Lee, N., (2019), Water Policy and Institution in the Republic of Korea, Asian Development Bank Institute, No. 985.

10. Asli, H.H., and Arabani, M., (2022), "Analysis of Strain and Failure of Asphalt Pavement. Computational Research Progress in Applied Science & Engineering", Transactions of Civil and Environmental Engineering, 8, 1–11. Article ID: 2250. https://doi.org/10.52547/crpase.8.1.2250.

11. Heino, O., and Takala, A., (2020), "Transformation of Urban Water Service Provision: Potential of Hybrid Systems", Public Works Management & Policy, 25(2), 151–166.

12. Asli, H.H., (2023), Modeling of Corrosion for Water System by Networked Sensors and the Internet of Things (IoT) in Compliance with Geography Information System (GIS). Sustainable Water Treatment and Ecosystem Protection Strategies. ISBN: 9781774915189. https://www.appleacademicpress.com/sustainable-water-treatment-and-ecosystem-protection-strategies-/9781774915189.

13. Kim, S., Devineni, N., Lall, U., and Soo Kim, H., (2017), Sustainable Development of Water Resources: Spatio-temporal Analysis of Water Stress in South Korea. MDPI.

14. Asli, H.H., Arabani, M., and Golpour, Y., (2020), Reclaimed Asphalt Pavement (RAP) Based on a Geospatial Information System (GIS). Slovak Journal of Civil Engineering, 28(2), 36–42. https://doi.org/10.2478/sjce-2020-0013. Slovak University of Technology in Bratislava, Slovak.

15. Hariri Asli, K., Hariri Asli, H., Motlaghzadeh, K., and Hariri Asli, K., (2013), "Numerical Techniques in Water Transmission", Frontiers of Engineering Mechanics Research (FEMR), August, 2(3), 56–62, ISSN: 2306–6016 (Online), ISSN: 2306–6024 (Print), published by the world academic publishing co., limited, Hong Kong, Corpus ID: 108917427. http://www.academicpub.org/femr/, https://api.semanticscholar.org/CorpusID:108917427.

16. Hariri Asli, K., and Hariri Asli, K., (2022), Isolated Pressure Zones Based on GIS as a Solution for Water Network Problems. Water Practice and Technology. https://doi.org/10.2166/wpt.2022.119.

17. BIPE 2015 FP2E/BIPE Report, (2015), Public Water and Wastewater Services in France Economic, Social and Environmental Data, Les Services Publics D'assainissement En France.

18. Bielsa, J., Cazcarro, J., Groot, E., and Sanchez Choliz, J., (2009), El Coste Financiero De La DMA. Tarifas Sobre El Uso Del Agua En Agricultura.

19. Hariri Asli, K., Hariri Asli, K., and Nazari, S., (2023), Computational Fluid Dynamics Analysis for Smart Control of Water Supply. Water Supply, 23(12). https://doi.org/10.2166/ws.2023.306.
20. Young Kim, H., et al., (2017), Water Resources Management in the Republic of Korea. Inter-American Development Bank-IDB.
21. Hariri Asli, K., and Hariri Asli, K., (2023), Minimum Night Flow (MNF) and Corrosion Control in Compliance with Internet of Things (IoT) for Water Systems. Water Practice and Technology. https://doi.org/10.2166/wpt.2023.012.
22. Waterworks Statistics, (2016), Each Year, Ministry of Environment.
23. Asli, Kih., and Asli, Kah., (2023), "Smart Water System and Internet of Things", Journal of Modern Industry and Manufacturing, 2, 5. https://doi.org/10.53964/jmim.2023005.
24. Kim, H.Y., et al., (2018), Water Resources Management in the Republic of Korea. Korea's Challenge to Flood & Drought with Multi-purpose Dam and Multi-regional Water Supply System.
25. Asli, Kih., and Asli, Kah., (2023), "Smart Heating, Ventilating, Air-conditioning and Refrigeration by Web-based Geographic Information System", Journal of Modern Industry and Manufacturing, 2, 6. https://doi.org/10.53964/jmim.2023006; https://www.innovationforever.com/article.jmim20230139.
26. Chamasemani, F.F., and Singh, Y.P., (2011), Multi-class Support Vector Machine (SVM) Classifiers—An Application in Hypothyroid Detection and Classification. 2011 Sixth International Conference on bio-inspired computing: theories and applications.

Optimization of Fixtures Unit Consumption by Intelligent Data Monitoring Method

7

Abstract

Today, there are various methods to collect water consumption data. Approximate estimates are the only method that was used in the past, while with the advancement of engineering tools and equipment, today more accurate measurements can be used to estimate the amount of consumption of each device. The final consumers of water include equipment in which low pressure water. This equipment includes bathrooms, toilets, washing machines, faucets, irrigation equipment, washing equipment, cooling equipment, and humidifying equipment. Water leakage from facilities can also be placed in the same group. It is important to accurately measure consumption in each of the water-consuming components. Knowing the amount of water consumed is one of the most essential information needed by urban water managers and planners. Investigating the efficiency of the water consumption reducer can lead to the modification of the design of the water consumption reducer. In this research, after the total consumption data was collected, the data were entered into the consumption analysis software. The effect of using a water consumption reducer on per capita consumption was investigated. The software determined the time, place, and amount of water consumed. The installation of consumption reducers on the Fixtures Unit reduced the per capita consumption by 78 Lit/day.

Keywords

Consumption reducer • Consumption per capita • Intelligent metering • Isolated area • Consumption modeling

7.1   Introduction

The final consumers of water include all the places that consume water. Most of the final consumers of water are:

- Household: small buildings and residential complexes
- Commercial-industrial: institutions, schools, government offices, parks
- Agriculture
- Energy production
- Other uses

7.1.1   Home Use

Among the final consumers of water, domestic consumers form the biggest sector. Therefore, water consumption by domestic consumers has always been the focus of researchers, planners, and water industry practitioners. Home consumption sectors are:

- Toilet
- Kitchen: cooking, eating, washing
- Bathroom: showers and bathtubs
- Washing machines: manual and automatic
- Dishwashers: manual and automatic
- Watering systems
- Cooling and humidifying: such as used in water coolers
- Leakage

7.1.2   Commercial and Industrial Uses

The second largest category of urban water consumers is commercial and industrial uses. This category of consumers uses water in various ways. For some professions, water is one of the main components of the activity. Due to the diversity and extent of water usage in these businesses, the final consumption of this sector cannot be accurately determined. Some of the types of water use include the following:

- Internal uses: toilets, faucets, showers, washing, cooking
- Expenses related to jobs: hairdressing salons, blacksmithing, greengrocers
- Cooling and humidifying: such as used in water coolers
- Leakage

7.1.3 Use of Consumption Data

Collecting consumption statistics and information has many uses, although they are often under the topics of water resources management and planning. Some of the applications of final consumption statistics are:

- Use in consumption forecasting models
- Use in demand management
- Planning
- Health studies
- Energy planning

7.1.4 Data Collection Methods

In general, it is possible to collect consumption data in the following ways:

- Estimate
- Estimation based on macro information of the network: Whenever the information is estimated based on the macro information of the network, the accuracy of the obtained information is also at the level of accuracy of the input information
- Estimation based on home measured data
- Estimation based on consumer audits: random audits of operating units and completion of questionnaires on their consumption levels can lead to the completion of information
- Direct measurement

Information estimation is often based on previous information. In this way, based on the previous observations of consumption information and also some indicators, a logical relationship is established between the information and the aforementioned indicators. Among the predictive models that have been developed in this regard, we can mention the linear and non-linear regression models of artificial neural networks, the fuzzy regression model, and the neuro-fuzzy model. All models are first trained with observational data and then analyzed with experimental data. If the output of the model has a slight difference with the actual observed value, it is accepted as a predictive model [1–5].

7.1.5 Data Collection Methods

The types of consumption analysis methods are: intelligent filtering of information, measuring the consumption of each component and using a daily diary.

7.2   Materials and Methods

Direct measurement of consumption of each component is the best method to deter-mine the share of each component. During the past years, various methods have been proposed for direct current measurement. Smart current measuring devices analyze the recorded data manually or by using software. An information recording device can also be connected to these devices.

In this research, the intelligent method of data collection and analysis was used to investigate the effect of using consumption reducers. By using this device, it is possible to record the amount of consumption of all faucets. In addition, by installing the trans-mitter on each of the current measuring devices installed on the faucet, the data can be transferred to the central receiver for recording in the computer. Such devices can be used to record and record data on an hourly, daily, and monthly basis. Flow tracking method has been accepted as a very efficient and reliable method. Currently, this method is used in many Non-Revenue Water (NRW) analysis projects.

7.2.1   Water Loss Calculation (7.1)

$$\text{Water Loss} = (\text{Quantity of water put in to supply}$$
$$- (\text{Non-domestic usage} + \text{Domestic consumption})) \qquad (7.1)$$

7.2.2   Non-Revenue Water (NRW) (7.2)

$$U = S - \left(M + \left(A * P_{\text{ps}}\right)\right) \qquad (7.2)$$

7.2.3   Non-Revenue Water (NRW) in the Distribution Network

Non-Revenue Water (NRW) in the water distribution network includes the following situations:

- Water losses due to fractures and incidents (visible leakage)
- Water losses due to small leaks (invisible leaks)
- Water losses from faucets and fittings
- Removal of fire extinguishers
- Washing the network through the available drain valves

Small leaks caused by faucets and fittings are considered to be part of small or invisible leaks. In this way, the minimum amount of water not taken into account due to faucets and fittings can be calculated. Since leaks caused by broken pipes and connections in the water distribution network are identified and fixed after receiving a leak report from residents, leaks caused by network failures and incidents are considered visible types of leaks. Regarding the leakage caused by the removal of fire hydrants, according to the investigations, this consumption is supplied through the wells that have been allocated for this purpose, and it has not been included in the calculations of the leakage from the network [6].

7.2.4 Intelligent Data Monitoring Method

The intelligent data monitoring method consists of a device for recording and transmitting water consumption data. The intelligent monitoring method can collect data once every 10s for cold water and once every 15s for hot water. The data recording device is connected to the output of the very accurate current measuring device. A very accurate current-measuring device is a device that can measure current once every 10s. Time intervals of 10s are a suitable and balanced amount in terms of data volume and desired accuracy [7–9].

The data related to the total water consumption are stored in a consumption analysis system. Then using Current consumption analysis software detects the relevant consumption type. For purposes that it is difficult or impossible to determine the type of consumption, an expert and skilled person, based on graphic outputs software can detect the type of consumption in the relevant hour. Then, the expert will teach the software model. So that the model after training can recognize types of consumption based on flow-time values. In the smart information monitoring method, there is no need to enter the houses to install the equipment. According to this method, water consumption data related to two weeks of measurement (about 120,000 data) can be analyzed in less than 2 h. In this process, the contribution of each consumption component is determined.

In general, the more the network data, the simpler and more accurate its analysis. One of the recommended ways to increase accuracy is to visit homes and check faucets by experts. So, if the amount of cost and time are not important factors, it is possible to check this method further. The advantages of this method are [10–15]:

• Only one current measuring device is needed, which can be easily installed.
• If there is a water meter in place, there is a need for additional piping or changes in the door.
• The consumer does not exist.
• There is no need to enter the houses.

The disadvantages of this method are:

- Expensive equipment for data recording, flow measurement, and software.
- In pressures less than 8 m of water, such as the points located after the pressure reducer valve (PRV), this method is not applicable.

Among the disadvantages of this method, the following points can be mentioned:

- The use of this method is suggested only in newly constructed buildings and before delivery to the customer.
- For large statistical populations, this method is not recommended.
- A lot of plumbing is needed to install the equipment.
- It is not suitable for use in low-pressure places.
- Using this method requires having a large number of flow meters.

The disadvantages of this method are more than its advantages. Due to the large number of meter installations in this method, the possibility of device leakage and damage increases [16].

7.2.5 Using a Daily Notebook

In this method, a notebook is given to each household. They are asked to note down the amount of consumption and the type of water consumed. For example, note the number of times you use the toilet or the duration of your bath [17–25].

7.3   Results

The present work investigated the effects of using consumption reducers on per capita consumption. To determine the amount of consumption of each fixture unit, this work was conducted based on statistics among 550 consumers. The results of the work in the first variant: without using the consumption reducer, and the second variant: using the consumption reducer for the Fixtures Unit, were compared with each other (Figs. 7.1, 7.2, 7.3 and 7.4). The results showed that the per capita consumption is 178 L per person per day (Table 7.1). Accurate measurement of the consumption of each of the components of consumption is very important to evaluate the methods of reducing consumption and their effectiveness. Statistics of final consumers are one of the most necessary components to improve consumption control methods. Knowing where and how much water is consumed is one of the most important information related to management that every urban water manager and planner should be aware of.

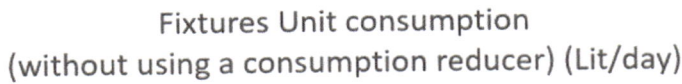

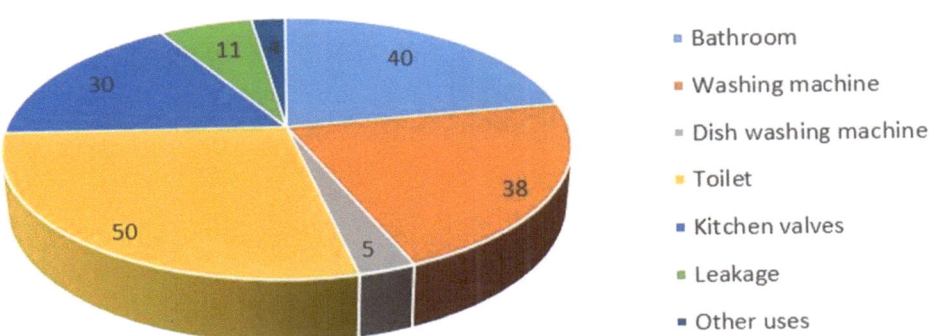

Fig. 7.1 Pie chart for changes in per capita water consumption without using consumption reducers for Fixtures Units consumption

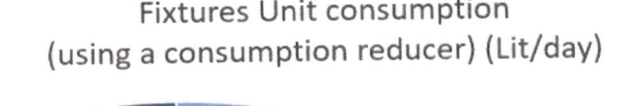

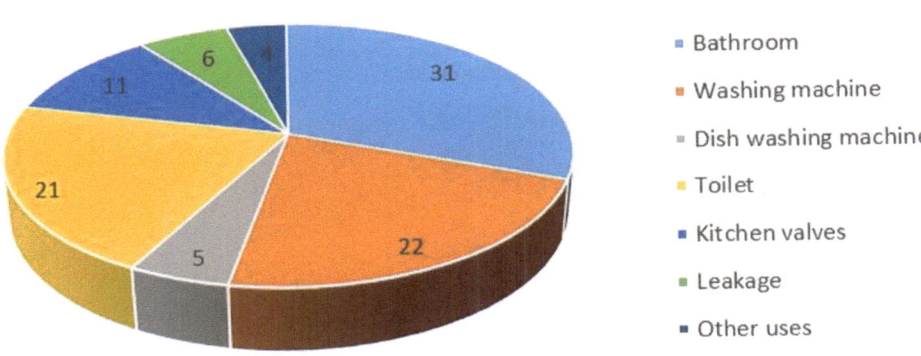

Fig. 7.2 Pie chart for changes in per capita water consumption due to using consumption reducers for Fixtures Units consumption

7.4   Conclusion

In this work, the consumption share of each of the consumption components was recorded by installing a flow meter on the outlets. The most important advantage of using this method was the direct measurement of the current in each of the outputs separately. A flow meter, flow recorder, and modem were installed on each of the consumption components,

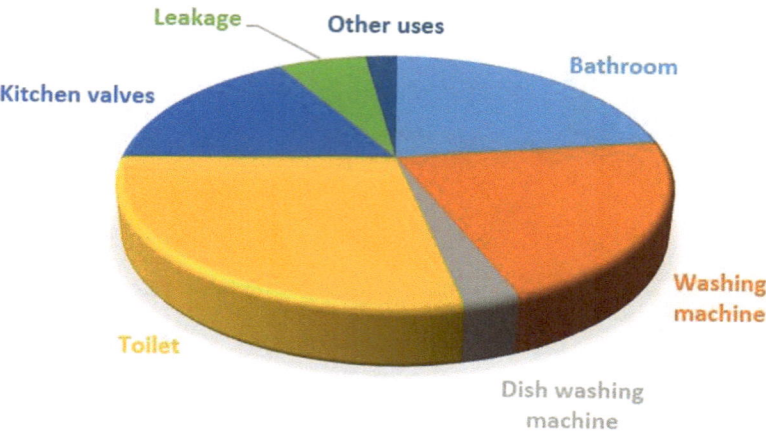

Fig. 7.3 Pie chart for changes in per capita water consumption without using consumption reducers for Fixtures Units consumption

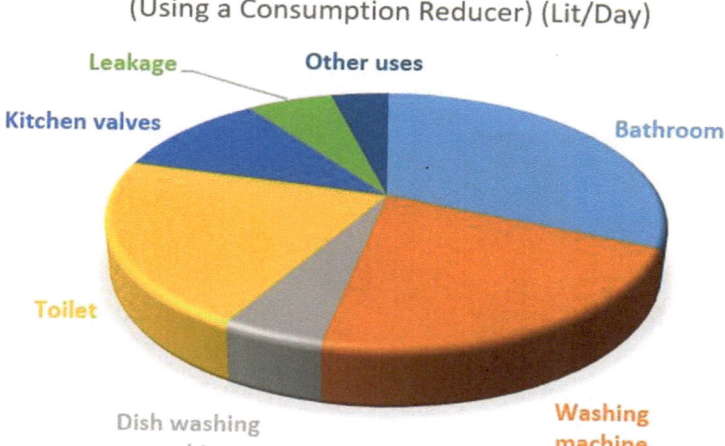

Fig. 7.4 Pie chart for changes in per capita water consumption due to using consumption reducers for Fixtures Units consumption

Table 7.1   Changes in per capita water consumption due to using consumption reducers for Fixtures Units consumption

Kind of fixtures unit	Fixtures unit consumption (without using a consumption reducer) (liters/day)	Fixtures unit consumption (using a consumption reducer) (liters/day)
Bathroom	40	31
Washing machine	38	22
Dish washing machine	5	5
Toilet	50	21
Kitchen valves	30	11
Leakage	11	6
Other uses	4	4
Total	178	100

such as faucets, toilets, bathrooms, and kitchens. Each modem sends information to the central system with its frequency. The outputs of this system include the name of the consuming part, the start time, the end time, the average flow rate, the peak flow rate, the duration of use, and the volume of water used. The results of the present research were as follows:

- The results of all consumption components were recorded.
- The results were divided into time, duration, and method of consumption.
- The ability to save data in computer files was provided.
- Installation of consumption reducers on Fixtures Unit reduced consumption by 78 Lit/ day.

The above studies showed that the per capita consumption is 178 L per person per day. Accurate measurement of the consumption of each of the consumption components is very important to evaluate the consumption reduction methods and their effectiveness. Statistics of final consumers are one of the most necessary components to improve consumption control methods. Knowing where and how much water is consumed is one of the most important information related to management that every urban water manager and planner should be aware of.

Suggestions for Future Research

This work investigated the variation in water consumption per capita due to the use of a consumption reducer. It can be subjected to future work on the water consumption study based on the Internet of Things (IoT).

Abbreviations

U Non-Revenue Water (NRW) for quantities of water including leakage

S Sum of all water inputs into a system

M Sum of all water accounted for by measure (metered supplies, domestic and non-domestic)

A Average domestic usage per capita of population

P_{ps} Population supplied (non-metered)

U Non-Revenue Water (NRW)

References

1. Creaco, E., and Walski, T., (2017), "Economic Analysis of Pressure Control for Leakage and Pipe Burst Reduction", Journal of Water Resources Planning and Management, 143(12).
2. Asli, H.H., and Hozori, A., (2021), "Non-Revenue Water (NRW) and 3d Hierarchical Model for Landslide", Larhyss Journaln 48, 189–210. http://larhyss.net/ojs/index.php/larhyss/article/view/810/810.
3. Calatrava, J.,and Garrido, A., (2010), Measuring Irrigation Subsidies in Spain: An Application of the GSI Method for Quantifying Subsidies. International Institute For sustainable development (IISD), Geneva.
4. Hariri Asli, H., and Nazari, S., (2021), "Water Age and Leakage in Reservoirs: Some Computational Aspects and Practical Hints", Larhyss Journal, 48, 151–167. http://larhyss.net/ojs/index.php/larhyss/article/view/808/807.
5. Bixioa, D., Thoeyea, C., De Koningb, J., Joksimovicb, D., Savicc, D., Wintgensd, T., and Melind, T., (2006), Wastewater Reuse in Europe, Presented at the International Conference on Integrated Concepts on Water Recycling, Wollongong, NSW, Australia, 14–17 February 2005.
6. Asli, H.H., (2023), Applications of Networked Sensors and the Internet of Things (IoT) for Water Treatment. Sustainable Water Treatment and Ecosystem Protection Strategies. ISBN: 9781774915189. https://www.appleacademicpress.com/sustainable-water-treatment-and-eco system-protection-strategies-/9781774915189.
7. European Academies, (2016), Indicators for a Circular Economy, EASAC policy report 30. European Environment Agency (EEA) Technical Report (2013) "Assessment of Cost Recovery Through Water Pricing". Publications Office, Luxembourg.
8. Asli, H.H., and Arabani, M., (2022), "Analysis of Strain and Failure of Asphalt Pavement. Computational Research Progress in Applied Science & Engineering", Transactions of Civil and Environmental Engineering, 8, 1–11. Article ID: 2250. https://doi.org/10.52547/crpase.8.1.2250.
9. Federal Ministry for the Environment, Nature Conservation and Nuclear Safety, (2011), The German Water Sector Policies and Experiences, Institut Für Umwelttechnik Und Management An Der Universität Witten/Herdecke Ggmbh.
10. Asli, H.H., (2023), Modeling of Corrosion for Water System by Networked Sensors and the Internet of Things (IoT) in Compliance with Geography Information System (GIS). Sustainable Water Treatment and Ecosystem Protection Strategies. ISBN: 9781774915189. https://www.appleacademicpress.com/sustainable-water-treatment-and-ecosystem-protection-strategies-/9781774915189.

11. Federal Statistical Office (Statistisches Bundesamt), (2016), Waste Management In Germany, Federal Environment Minstry (BMUB).
12. Asli, H.H., Arabani, M., and Golpour, Y., (2020), Reclaimed Asphalt Pavement (RAP) Based on a Geospatial Information System (GIS). Slovak Journal of Civil Engineering, 28(2), 36–42. https://doi.org/10.2478/sjce-2020-0013. Slovak University of Technology in Bratislava, Slovak.
13. Haselbach, L., Adesina, M., and Muppavarapu, N., et al., (2023), "Spatially Estimating Fooding Depths from Damage Reports", Natural Hazards, 117, 1633–1645. https://doi.org/10.1007/s11 069-023-05921-2.
14. Weidner, J., Collins, J., Benitez, M., Adesina, M., and Lozoya, C., (2019), Development of a Robust Framework for Assessing Bridge Performance Using a Multiple Model Approach, University of Texas at El Paso. Department of Civil Engineering, report number: CAIT-UTC-NC39. https://rosap.ntl.bts.gov/view/dot/48948.
15. Hariri Asli, K., Hariri Asli, H., Motlaghzadeh, K., and Hariri Asli, K., (2013), "Numerical Techniques in Water Transmission", Frontiers of Engineering Mechanics research (FEMR), August, 2(3), 56–62, ISSN: 2306–6016 (Online), ISSN: 2306–6024 (Print), published by the world academic publishing co., limited, Hong Kong, Corpus ID: 108917427. http://www.academicpub.org/femr/; https://api.semanticscholar.org/CorpusID:108917427.
16. Hariri Asli, K., and Hariri Asli, K., (2022), Isolated Pressure Zones Based on GIS as a Solution for Water Network Problems. Water Practice and Technology. https://doi.org/10.2166/wpt.202 2.119.
17. France's Environmental Policy is Proactive and Ambitious, As Exemplified In (2015) By the Energy Transition for Green Growth Act and the Paris Agreement at cop21, and in 2016 by the Draft Law on Biodiversity.
18. Hariri Asli, K., Hariri Asli, K., and Nazari, S., (2023), Computational Fluid Dynamics Analysis for Smart Control of Water Supply. Water Supply. https://doi.org/10.2166/ws.2023.306.
19. OECD, (2014), Environmental Taxation a Auide for Policy Makers, Guide is Based on the OECD Recently Issued Book Taxation, Innovation and the Environment.
20. Ruiz-Villaverde, A., and Garcia-Rubio, M.A., (2017), Public Participation in European Water Management: From Theory to Practice. Water Resources Management.
21. Hariri Asli, K., and Hariri Asli, K., (2023), Minimum Night Flow (MNF) and Corrosion Control in Compliance with Internet of Things (IoT) for Water Systems. Water Practice and Technology. https://doi.org/10.2166/wpt.2023.012.
22. Asli, Kih., and Asli, Kah., (2023), "Smart Water System and Internet of Things", Journal of Modern Industry and Manufacturing, 2(5). https://doi.org/10.53964/jmim.2023005; https://www.innovationforever.com/article.jmim20230111.
23. Asli, Kih., and Asli, Kah., (2023), "Smart Heating, Ventilating, Air-conditioning and Refrigeration by Web-based Geographic Information System", Journal of Modern Industry and Manufacturing, 2, 6. https://doi.org/10.53964/jmim.2023006; https://www.innovationforever.com/article.jmim20230139.
24. Soes-SSP, Water Survey, (2008), French Ministry of Health, ARS, SISE-Eaux Database.
25. The Monitoring and Statistics Directorate (Soes), (2017), 10 Key Indicators for Monitoring the Circular Economy, Ministry of the Environment, Energy and Marine Affairs, in Charge of International Relations on Climate Change, France.

Remote Sensing (RS) and Domestic Consumption Improvers Equipment

8

Abstract

Investigating different consumption management options to prevent water loss should include all types of analysis including cost–benefit analysis. The cost–benefit analysis includes current production cost, income level, profitability of project implementation, and technical, economic, and operational comparison of different projects. In the conditions of lack of financial resources, it is necessary to perform a cost–benefit analysis. To prevent water wastage, household consumers can control the uses outside their home, such as gardening, and car washing, through different methods. These methods can be physical using tools or natural. For example, planting plants with low water consumption and appropriate to the climatic conditions of the region is a natural method. Using water sprinklers as a physical method can improve the efficiency of using water resources. In general, the use of consumption-reducing devices should have the same function as the usual sanitary facility equipment. The consumption-reducing installation should not reduce the work efficiency of the existing sanitary facility system. These devices should even be able to increase the efficiency of daily activities while using sanitary facilities. The current work, has investigated the effect of installing consumption improvement tools on improving efficiency and saving in the use of water resources. A flow meter, flow recorder (data logger), and modem were installed on each of the consumption components, such as faucets, toilets, bathrooms, and kitchens. Each modem sends information through Remote Sensing (RS) to the central system with its frequency. The effects of using a water consumption reducer on per capita consumption also was investigated. The installation of consumption reducers on the Fixtures Units led to a 43.82% saving in drinking water.

Keywords

Consumption reducer • Consumption per capita • Sanitary facilities • Isolated area •
Drinking water

8.1   Introduction

In the twenty-first century, the population of the earth has increased several times due
to the improvement of public health. The world population in 2023 is about 8 billion
people. The demand for freshwater has increased many times. If the current trend con-
tinues, by 2050 water resources will meet the daily needs of 50 L of drinking water per
person. Today, the use of useful tools and methods for the optimal and effective use of
water is recommended by all authorities and organizations. One of the primary ways to
conserve water is to use correct water. This can be done through the modification of the
consumption pattern, public education, the use of advanced tools and equipment for consumption reduction, reconsideration of the design methods, improvement of the operation
of the leakage control network, as well as the use of social and economic tools (setting
tariffs appropriate) to take place. Using tools such as consumption reducers is one of the
common and effective ways to conserve water resources. In the domestic sector of water
consumption, consumer units can try to reduce water consumption by using reducing
tools. These tools generally include a flow rate reducer, pressure reducer, water output
speed increaser (sprayer), tank volume reducer, flow meter, flow purification system and
water reuse tools. Flow measurement causes the consumer to be sensitive to the amount
of consumption of each activity and the price of water corresponding to that activity.
The use of consumption-reducing tools can ultimately lead to the encouragement of the
consumer to reduce water consumption. Flow control tools, Physically, cause obstacles
and limitations in the volume of water outflow. The desired control can also be done by
actually reducing the flow rate by adding air to the water (occupying part of the water
volume with air). In systems with reduced water pressure, less water will be distributed
to the consumer at a certain time. As a result, consumption will decrease. Increasing the
speed of water exit from the outlet creates more efficiency in some activities such as
washing the car and washing the toilets. This is possible by installing a flow-level output
controller. For some areas that suffer from severe water shortage, it is logical to install
a water consumption improver to prevent water wastage. Of course, such devices are
usually not recommended except for exceptional cases and areas. Also, in cases such as
using siphons, reducing the useful volume of the siphon tank is one of the ways to save
water. Consumption-reducing tools have the same standards as normal tools. These tools
are effective in the amount of water consumption. In most of the countries of the world,
there are specific standards defined and available for building sanitary faucets that match
the conditions of each country.

After being implemented for many years, the above standards have undergone changes and transformations after revision and continue to evolve. The above standard includes a collection of information on technical specifications and various tests of different types of faucets, overhead showers, bathroom shower hoses, and lever mixer tap. The implementation of some cases is non-compulsory and some are mandatory. The above standard includes the following items [1–5].

- Sanitary faucets
- Sanitary faucets
- Mechanical lever mixed valves
- Bathroom shower head
- Bathroom shower hose

To prevent water wastage, home consumers can control the uses outside their homes, such as gardening, car washing, through different methods. These methods can be physical using tools or natural. For example, planting low-consumption plants suitable for the climatic conditions of the region is a natural method. Using water sprinklers as a physical method can improve the efficiency of using water resources.

The current research has investigated the effect of installing consumption improvement tools on improving efficiency and saving in the use of water resources.

8.2   Materials and Methods

Regarding the maintenance of domestic water sources, the use of all tools such as the use of consumption reducers, the establishment of appropriate tariffs to encourage consumption reduction, the calculation of losses, and the improvement of the water leakage situation are emphasized. In the field of household consumption, valuable studies have been conducted in different parts of the world.

8.2.1   Water Loss Calculation (8.1)

$$\text{Water Loss} = (\text{Quantity of water put in to supply} \\ - (\text{Non-domestic usage} + \text{Domestic consumption})) \qquad (8.1)$$

8.2.2   Non-Revenue Water (NRW) (8.2)

$$U = S - \left(M + \left(A * P_{ps}\right)\right) \tag{8.2}$$

8.2.3   Non-Revenue Water (NRW) in the Per Capita of Domestic Usage

Some researchers have investigated the amount of consumption of each component of household consumption. The difference between the numbers is based on the population sample, place, and time of selection. About 65% of water consumption is in the toilet and bathroom. Therefore, reducing the consumption of the toilet and bathroom is one of the important goals of reducing consumption in water-scarce countries. Because reducing consumption in these sectors can have major effects on reducing the consumption of the entire household sector. During the implementation of consumption management, the amount of consumption of each household sector is investigated and determined. Weather conditions, climate, culture, and the amount of consumption of different components are different. In the area of consumption outside the home, the main consumption is watering the lawn and washing the car.

8.2.4   Reducing Devices Toilets

The use of these devices can greatly affect the total volume of toilet consumption. It is common for this type of toilet to have specific standards for the production and use of reducing devices for this type of toilet It is also mandatory. In the past, it was common to use toilets with a tank volume of 21 L. The use of these toilets leads to the consumption of 49,140 L of fresh water to dispose of 620 L of feces per person in a year. During the last 50 years, by reviewing the design of such toilets, 13-L toilets with the same efficiency as 21-L toilets were released to the market. In this way, there was a huge reduction in the volume of freshwater used (about 36%) in the production of wastewater. The technology used in very low-consumption toilets is the use of trapped air pressure to drive water under higher pressure into the toilet bowl. In this way, using a smaller volume of water and using air pressure, the toilet bowl was cleaned. After each use of the tank, water flows into the tank through the hose. The water flow will continue until the air pressure and water pressure reach equilibrium. In this way, the air in the tank is compressed and during washing, it will help the flow of water that flows by gravity. The efficiency of these toilets depends on the water pressure. Along with the existence of reducing devices, there are also recommendations regarding the necessity of repairs, maintenance and how to use siphons, which in turn can be effective in reducing the volume of water consumed by siphons. One of the most important things is checking the existence of leaks in siphons. This can be done by listening to the sound of the leak. To check the existence of leakage, you can pour some colored substance into the siphon tank. After 15 min, if the concentration of the colored substance decreases, it means that there is a leak. Otherwise, there is no leakage [6, 7].

8.2.5   Reducing Devices that Can Be Used in the Bathroom

Bathroom water consumption is the second most used water in residential buildings after toilets. The volume of water used in the bathroom depends on the flow rate of the shower head and the duration of using the bathroom. Therefore, the reduction of bathing time and the reduction of the discharge from the shower head should be considered. Regarding the reduction of bathing time, the culture of using the bathroom should be corrected by informing the people. Reducing the discharge of head showers is possible by improving the physical performance of head showers.

8.2.6   Standard for Showers

The purpose of compiling the bathroom shower head standard is to determine the classification, specifications, materials, chemical and sanitary properties, external surface conditions, and the quality of its coating. The standard for overhead showers includes dimensional features, waterproofing specifications, mechanical specifications, hydraulic specifications, acoustic specifications, maintenance method, rotating connection, marking, and how to test these features in the bathroom shower head. This standard is used for overhead showers and hand showers for washing. At the same time, they are applicable for equipping and completing sanitary faucets used in bathrooms and showers.

The purpose of developing the bathroom shower hose standard is to determine the following features in the bathroom shower hose:

- Determination of classification
- Specifications
- Materials
- Chemical properties
- Hygiene, external surface conditions, and the quality of its coating
- Dimensional characteristics, waterproofing characteristics
- Mechanical specifications, hydraulic specifications
- Audio specifications
- Preservation method
- Rotary joint

- Marking and how to test

The reduction of the output flow is done by installing water heaters that reduce consumption or by installing a flow regulator in the pipeline. This standard is about the hoses that are used as the connection of building plumbing and sanitary appliances. Hoses that are part of the plumbing circuit. But for example, the hose connecting the dishwasher faucet and the plumbing circuit is not used. The bathroom shower hose must be installed after turning off the faucet. The output flow rate can be reduced by installing water-reducing shower heads or by installing a flow regulator inside the pipeline [8–10].

8.2.7   Consumption-Reducing Shower Heads

The consumption of common showers without using any type of reducer is about 18 L per minute. Including 5 min for each person to take a shower, about 90 L of clean water enters the sewage network with each shower. In order to reduce consumption, new shower heads were produced with the aim of replacing old shower heads and reducing consumption. In the design of low-consumption showers, by limiting the outlet orifice, the flow rate was reduced to 10 L per minute. Using this type of showerhead led to 44% savings in water consumption. By using air spray technology and mixing water and air, the water consumption of showers can be reduced to 5 L per minute. Using these types of showers for a 5-min bath saves nearly 70 L compared to normal showers. Of course, water consumption has been reduced according to the customer's satisfaction with the product and its use. This issue can be found in the repeated reports of the relevant organizations. In such products, the combination of air pressure and gravitational force leads to an increase in the speed of the water coming out of the showers.

In this way, the efficiency of using shower heads increases. The use of this type of shower head reduced the savings by 60% compared to the use of head shower heads without a reducer. It also caused 30% savings compared to low consumption showers [11–13].

8.2.8   Reducing Devices that Can Be Used in Faucets

About 5 percent of the total volume of domestic water consumption is used in the toilet. Used water enters the sewage system. This amount may be small compared to the amount of water used in the toilet and bathroom, but by improving the consumption of the same amount of water on a large scale, huge savings can be made. Saving the amount of water used in faucets will also save energy. The consumption of ordinary faucets is equal to 10 to 19 L per minute. The consumption of milk machines with new technologies is about 9 L per minute per day. The current work investigated the use of flow limiting valves as the desired technology and the most basic ways to reduce consumption. Nowadays, the use of aerators is significantly more efficient than other methods [14].

8.2.9 Air Valves

Aerated valves cause water and air to mix while the water exits the pipe opening. This act in the fact that It reduces both the intensity of the outgoing water and the dispersion of water, increasing the efficiency and performance It becomes milk. Therefore, at the same time, saving water consumption and increasing efficiency happen. Aerators do not change the amount of water needed to fill a container or dishwashing bowl. But with the use of aerators, the water required for rinsing can be significantly reduced [15].

8.2.10 Reducing Consumption in the Washing Machine

In the houses where there is a washing machine, approximately 20% of the water consumption is dedicated to the washing machine. Saving in consumption is achieved by using different ways such as adjusting the washing machine load, adjusting the detergent tank, and water level, and designing the washing machine [16].

8.2.11 Pressure Reducers

The pressure of the water distribution system is one of the most important factors in determining the output flow rate. Home distribution system pressure varies between 20 and 40 m of water. More outlet pressure causes more water to come out [17–25].

8.3   Results

In this work, the most important issue after examining the economy of reducers was the analysis of the ease of use and the investigation of possible limitations caused by their use. The use of reducing devices in general should not cause any restrictions on the consumer. The idea is that the consumer should have the same feeling as using normal devices. Of course, this feeling is different for different consumers. Its average should be considered as a standard. Before starting the work, the data of each reducer was checked regarding the approximate amount of reduction in reducer consumption. Among the various options, with the awareness of the amount of reduction in consumption due to the use of the reducer, the price of water, and the life of the device, the correct choice of the reducer was made. After the project became operational, to evaluate the success of the project, the

Table 8.1 Changes in water consumption per capita (with and without consumption management variants)

Kind of Fixtures Unit	Water consumption per capita (without consumption management) (liters/day)	Water consumption per capita (with consumption management) (liters/day)	Water saving (%)
Bathroom	40	31	22.5
Washing machine	38	22	42.11
Dish washing machine	5	5	0
Toilet	50	21	58
Kitchen valves	30	11	63.34
Leakage	11	6	45.46
Other uses	4	4	0
Total	178	100	43.82

Table 8.2 Water saving (with and without consumption management variants)

No.	Kind of consumption	Water saving (%)
1	Bathroom	22.5
2	Washing machine	42.11
3	Dish washing machine	0
4	Toilet	58
5	Kitchen valves	63.34
6	Leakage	45.46
7	Other uses	0
	Total	43.82

effectiveness of the use of reducers was investigated. In this way, a specific strategy related to the extension of the implementation of the project to other areas or possible reforms in the way the project is implemented was formulated. The practical study revealed many factors that need correction these factors include cultural factors, how to use and the level of belief in the use of reducers, economic factors, purchasing power, and many other factors that can be correctly identified in the work stage. Another thing that was paid attention to in this work was investigating the efficiency of water consumption-reducing devices (Tables 8.1, 8.2, Figs. 8.1, 8.2). Consumption reducers should not cause unpleasant feelings in the consumers. For example, the use of aerators at the head of the toilet hose will reduce the speed of the water, reduce the cleaning power, and as a result, reduce the efficiency of using this device. On the contrary, the use of spray accelerators will not only reduce consumption in toilets but will also increase consumer satisfaction.

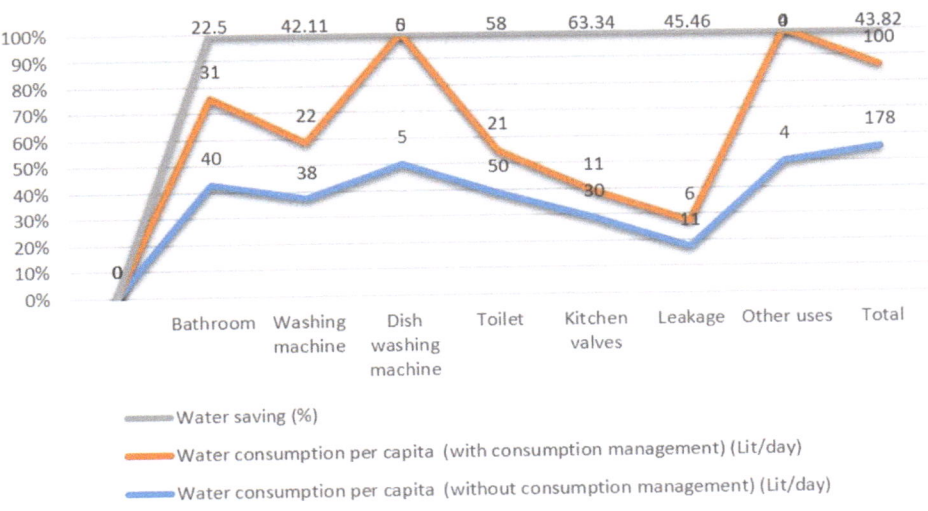

Fig. 8.1 Changes in water consumption per capita (with and without consumption management variants)

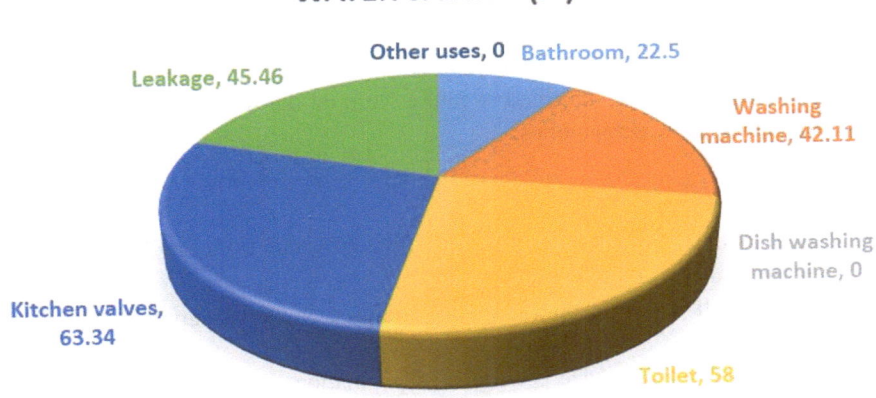

Fig. 8.2 Pie chart for water saving (with and without consumption management variants)

8.4   Conclusion

In this work, the consumption share of each of the consumption components was recorded by installing a flow meter on the discharge line. The most important advantage of using this method was the direct measurement of the current in each of the outputs separately.

A flow meter, flow recorder (data logger), and modem were installed on each of the consumption components, such as faucets, toilets, bathrooms, and kitchens. Each modem sends information through remote sensing (RS) to the central system with its frequency. The outputs of this system include the name of the consuming part, the start time, the end time, the average flow rate, the peak flow rate, the duration of use, and the volume of water used. The results of the present work were as follows:

- The results of all consumption components were recorded.
- The results were divided into time, duration, and method of consumption.
- The ability to save data in computer files was provided.
- Installation of consumption reducers on Fixtures Unit reduced consumption by 78 Lit/day.

The above studies showed that the per capita consumption is 178 L per person per day. Accurate measurement of the consumption of each of the consumption components is very important to evaluate the consumption reduction methods and their effectiveness. Statistics of final consumers are one of the most necessary components to improve consumption control methods. Knowing where and how much water is consumed is one of the most important information related to management that every urban water manager and planner should be aware of.

Suggestions for Future Research

This work investigated the variation of water saving (with and without consumption management variants). It can be subjected to future work on the water-saving study based on the smart metering of water supplies.

Abbreviations

U	Non-Revenue Water (NRW) for quantities of water including leakage
S	Sum of all water inputs into a system
M	Sum of all water accounted for by measure (metered supplies, domestic and non-domestic)
A	Average domestic usage per capita of population
P_{ps}	Population supplied (non-metered)
U	Non-Revenue Water (NRW)

References

1. Hu, C., Li, M., Zeng, D., and Guo, S., (2018), "A Survey on Sensor Placement for Contamination Detection in Water Distribution Systems", Wireless Networks, 24(2), 647–661.https://doi.org/10.1007/s11276-016-1358-0. Springer, US.

2. Ohar, Z., Lahav, O., and Ostfeld, A., (2015), "Optimal Sensor Placement for Detecting Organophosphate Intrusions into Water Distribution Systems", Water Research, 73, 193–203.https://doi.org/10.1016/j.watres.2015.01.024. Elsevier Ltd.

3. Asli, H.H., and Hozori, A., (2021), "Non-Revenue Water (NRW) and 3d Hierarchical Model for Landslide", Larhyss Journal, 48, 189–210. http://larhyss.net/ojs/index.php/larhyss/article/view/810/810.

4. Veilleux, J., and Dinar, S., (2019), "A Global Analysis of Water-related Terrorism, 1970–2016 a Global Analysis of Water-related Terrorism, 1970–2016", Terror Polit Violence, 0(0), 1–26. https://doi.org/10.1080/09546553.2019.1599863. Routledge.

5. Asli, H.H., and Nazari, S., (2021), "Water Age and Leakage in Reservoirs: Some Computational Aspects and Practical Hints", Larhyss Journal, 48, 151–167. http://larhyss.net/ojs/index.php/larhyss/article/view/808/807.

6. Wang, S., Basten, C.J., and Zeng, Z. Windows QTL Cartographer, (2012), Department of Statistics, North Carolina State University, Raleigh, NC, USA https://brcwebportal.cos.ncsu.edu/qtlcart/WQTLCart.htm.

7. Asli, H.H., (2023), Applications of Networked Sensors and the Internet of Things (IoT) for Water Treatment. Sustainable Water Treatment and Ecosystem Protection Strategies. ISBN: 9781774915189.https://www.appleacademicpress.com/sustainable-water-treatment-and-ecosystem-protection-strategies-/9781774915189.

8. Asli, H.H., and Arabani, M., (2022), "Analysis of Strain and Failure of Asphalt Pavement. Computational Research Progress in Applied Science & Engineering", Transactions of Civil and Environmental Engineering, 8, 1–11. Article ID: 2250. https://doi.org/10.52547/crpase.8.1.2250.

9. Rathi, S., Gupta, R., and Kamble, S., (2016), "Risk Based Analysis for Contamination Event Selection and Optimal Sensor Placement for Intermittent Water Distribution Network Security", Water Resources Management, 30, 2671–2685. https://doi.org/10.1007/s11269-016-1309-7.

10. Asli, H.H., (2023), Modeling of Corrosion for Water System by Networked Sensors and the Internet of Things (IoT) in Compliance with Geography Information System (GIS). Sustainable Water Treatment and Ecosystem Protection Strategies. ISBN: 9781774915189. https://www.appleacademicpress.com/sustainable-water-treatment-and-ecosystem-protection-strategies-/9781774915189.

11. Preis, A., and Ostfeld, A., (2008), "Multiobjective Contaminant Sensor Network Design for Water Distribution Systems", Journal of Water Resources Planning and Management © ASCE /, (August), pp. 366–377.

12. Asli, H.H., Arabani, M., and Golpour, Y., (2020), Reclaimed Asphalt Pavement (RAP) Based on a Geospatial Information System (GIS). Slovak Journal of Civil Engineering, 28(2), 36–42. https://doi.org/10.2478/sjce-2020-0013. Slovak University of Technology in Bratislava, Slovak.

13. Ostfeld, A., Uber, J.G., Salomons, E., Berry, J.W., Hart, W.E., Phillips, C.A., Watson, J.P., Dorini, G., Jonkergouw, P., Kapelan, Z., and Pierro, F., (2008), "The Battle of the Water Sensor Networks BWSN ... A Design Challenge for Engineers and Algorithms", Journal of Water Resources Planning and Management, 134(6), 556–568.

14. Hariri Asli, K., Hariri Asli, H., Motlaghzadeh, K., and Hariri Asli, K., (2013), "Numerical Techniques in Water Transmission", Frontiers of Engineering Mechanics Research (FEMR), August,

2(3), 56–62, ISSN: 2306–6016 (Online), ISSN: 2306–6024 (Print), published by the world academic publishing co., limited, Hong Kong, Corpus ID: 108917427. http://www.academicpub.org/femr/; https://api.semanticscholar.org/CorpusID:108917427.

15. He, G., Zhang, T., Zheng, F., and Zhang, Q., (2018), "An Efficient Multi-objective Optimization Method for Water Quality Sensor Placement Within Water Distribution Systems Considering Contamination Probability Variations", Water Research, 143, 165–175.https://doi.org/10.1016/j.watres.2018.06.041. Elsevier Ltd.

16. Hariri Asli, K., and Hariri Asli, K., (2022), Isolated Pressure Zones Based on GIS as a Solution for Water Network Problems. Water Practice and Technology. https://doi.org/10.2166/wpt.2022.119.

17. Hariri Asli, K., Hariri Asli, K., and Nazari, S., (2023), "Computational Fluid Dynamics Analysis for Smart Control of Water Supply", Water Supply, 23(12). https://doi.org/10.2166/ws.2023.306.

18. Naserizade, S.S., Nikoo, M.R., and Montaseri, H., (2018), "A Risk-based Multi-objective Model for Optimal Placement of Sensors in Water Distribution System", Journal of Hydrology, 557, 147–159. https://doi.org/10.1016/j.jhydrol.2017.12.028. Elsevier B.V.

19. Haselbach, L., Adesina, M., and Muppavarapu, N., et al., (2023), "Spatially Estimating Flooding Depths from Damage Reports", Natural Hazards, 117, 1633–1645. https://doi.org/10.1007/s11069-023-05921-2.

20. Weidner, J., Collins, J., Benitez, M., Adesina, M., and Lozoya, C., (2019), Development of a Robust Framework for Assessing Bridge Performance Using a Multiple Model Approach, University of Texas at El Paso. Department of Civil Engineering, report number: CAIT-UTC-NC39. https://rosap.ntl.bts.gov/view/dot/48948.

21. Wackerbauer, J., (2009), The Water Sector in Germany, Centre International De Recherches Et Information Sur Economie Publique, Sociale Et Coopérative (CIRISE).

22. Hariri Asli, K., and Hariri Asli, K. (2023) Minimum Night Flow (MNF) and Corrosion Control in Compliance with Internet of Things (IoT) for Water Systems. Water Practice and Technology. https://doi.org/10.2166/wpt.2023.012.

23. Asli, Kih., and Asli, Kah., (2023), "Smart Water System and Internet of Things", Journal of Modern Industry and Manufacturing, 2, 5. https://doi.org/10.53964/jmim.2023005; https://www.innovationforever.com/article.jmim20230111.

24. Asli, Kih., and Asli, Kah., (2023), "Smart Heating, Ventilating, Air-conditioning and Refrigeration by Web-based Geographic Information System", Journal of Modern Industry and Manufacturing, 2, 6. https://doi.org/10.53964/jmim.2023006; https://www.innovationforever.com/article.jmim20230139.

25. The United Nations World Water Development, (2010), Water in a Changing World. UNESCO Publishing.